Principles of Thermal Analysis and Calorimetry

2nd Edition

Principles of Thermal Analysis and Calorimetry

2nd Edition

Edited by

Simon Gaisford

University College London, UK
Email: s.gaisford@ucl.ac.uk

Vicky Kett

Queen's University Belfast, UK
Email: v.kett@qub.ac.uk

Peter Haines

Oakland Analytical Services, Farnham, UK
Email: phaines875@virginmedia.com

THE QUEEN'S AWARDS
FOR ENTERPRISE:
INTERNATIONAL TRADE
2013

Print ISBN: 978-1-78262-051-8

A catalogue record for this book is available from the British Library

Published by The Royal Society of Chemistry,
Thomas Graham House, Science Park, Milton Road,
Cambridge CB4 0WF, UK

Registered Charity Number 207890

Visit our website at www.rsc.org/books

Printed in the United Kingdom by CPI Group (UK) Ltd, Croydon, CR0 4YY, UK

Preface

The Thermal Methods Group (TMG), part of the Royal Society of Chemistry (RSC), exists to promote the education, awareness and application of thermal and calorimetric methods in the UK. Over the years, many illustrious names in the thermal analysis field have served on the TMG committee and 2015 is a special year as it marks the 50th anniversary of its founding. As part of the anniversary celebrations, the current TMG Committee decided to produce a revised and updated version of its textbook, *Principles of Thermal Analysis and Calorimetry*. The first version, published in 2001, provided an excellent introduction to thermal analysis and calorimetric techniques, but the past decade has seen the development of better technology and new techniques. This revised text therefore seeks to address these new developments, in addition to providing the background to the core techniques as in the first edition.

The focus of this text is on the underlying theory and operating principles of each technique. Each chapter has been written by an expert in the field and has been reviewed by the Committee. A review of thermal analysis nomenclature, as approved by the International Union of Pure and Applied Chemistry, is provided before discussion of the individual techniques. In addition to the well-established techniques of differential scanning calorimetry (DSC), isothermal calorimetry, thermogravimetric analysis and mechanical analysis, this text discusses a wider range of techniques such as dynamic vapour sorption, dielectric thermal analysis, modulated-temperature DSC, sample-controlled thermal analysis and hazard calorimetry. Some recent, and very exciting, new approaches are also introduced, such as modifications to localised thermal analysis and hyphenated techniques.

Principles of Thermal Analysis and Calorimetry: 2nd Edition
Edited by Simon Gaisford, Vicky Kett and Peter Haines
© The Royal Society of Chemistry 2016
Published by the Royal Society of Chemistry, www.rsc.org

We hope, therefore, that you find the book interesting, informative and stimulating, and it helps you in starting to use thermal methods or in improving your current analyses. We also hope that you will join us in promoting thermal methods, through word-of-mouth or by contributing to the work of the TMG, either by attending our annual meetings and training courses, becoming a member or joining the Committee. Further details can be found at our web-site.

Simon, Vicky and Peter
www.thermalmethodsgroup.org.uk

Contents

Principles of Thermal Analysis and Calorimetry: 2nd Edition
Edited by Simon Gaisford, Vicky Kett and Peter Haines
© The Royal Society of Chemistry 2016
Published by the Royal Society of Chemistry, www.rsc.org

2 Thermal Analysis Nomenclature 14

Trevor Lever

3 Thermogravimetry 18

Vicky L. Kett and Duncan M. Price

4 Dynamic Vapour Sorption 47

Nicole Hunter

5 Differential Scanning Calorimetry 67

Paul Gabbott and Tim Mann

6 Modulated Temperature Differential Scanning Calorimetry 104
Vicky Kett

7 Isothermal Microcalorimetry 123
Simon Gaisford

8 Isothermal Reaction Calorimetry and Adiabatic Calorimetry 146
Ian Priestley

9 Thermomechanical, Dynamic Mechanical and Dielectric Methods   164
John C. Duncan and Duncan M. Price

10 Simultaneous Thermal Techniques   214
Ian J. Scowen and Richard Telford

11 Sample Controlled Thermal Analysis   232
G. M. B. Parkes and E. L. Charsley

1 TMG History

Edward L. Charsley,*[a] Peter J. Haines[b] and Fred W. Wilburn[c]

[a] IPOS, School of Applied Sciences, University of Huddersfield, Queensgate, Huddersfield HD1 3 DH, UK; [b] Oakland Analytical Services, 38 Oakland Avenue, Farnham, Surrey GU9 9DX, UK; [c] Southport, UK
*Email: e.l.charsley@hud.ac.uk

1.1 Introduction

In 2015, the Thermal Methods Group (TMG), an interest group of the Royal Society of Chemistry (RSC), entered its 50th year. It seems appropriate therefore to not only look back to its origins but also to review briefly its activities since its formation and its current role in the field of thermal analysis and calorimetry. For readers interested in a more detailed account of the formation of the Group, there is an excellent paper covering the first twenty one years of the TMG's history.[1]

1.2 The Early Years

The 1960s were an exciting time for thermal analysis. The field was growing rapidly and the range of commercially available equipment was expanding, including the introduction of both power-compensated and heat flux differential scanning calorimeters. In addition, a significant number of thermal analysts were engaged in building their own equipment, particularly in the field of simultaneous thermogravimetry–differential thermal analysis.

Principles of Thermal Analysis and Calorimetry: 2nd Edition
Edited by Simon Gaisford, Vicky Kett and Peter Haines
© The Royal Society of Chemistry 2016
Published by the Royal Society of Chemistry, www.rsc.org

At that time, there was no forum in the UK where thermal analysts could meet so, in 1964, John Redfern and Cyril Keattch decided to investigate the possibility of forming a group by holding a thermal analysis symposium at the Battersea College of Technology (now the University of Surrey). The response exceeded the organisers' expectations and some 100 delegates attended the meeting. The discussions that took place during the technical sessions, and at the buffet supper afterwards, enthusiastically endorsed the idea of formally establishing a thermal analysis group.

This aim was realised on the 24th February 1965 when the UK Thermal Analysis Group was formed as part of the Society for Analytical Chemistry. The inaugural meeting took place on the evening of May 28th 1965 at the Chemical Society, Burlington House, Piccadilly, where the new Committee was confirmed. The Group's first Chairman was Robert Mackenzie with John Redfern as Vice-Chairman, Cyril Keattch as Secretary and Alan Hodgson as Treasurer. The 1st AGM was held in December 1965 and the Group was well and truly launched with its first two-day meeting in April 1966. This consisted of a visit to the laboratories of Pilkington Bros. Ltd on the first day and a meeting at the Royal College of Advanced Technology, Salford (now the University of Salford), on the second day on the "Characterisation of Residues after Thermal Treatment".

In 1972, the name of the Group was changed to the Thermal Methods Group to reflect the wide range of techniques falling within its scope. Following the merger of the Society of Analytical Chemistry and the Chemical Society in 1975, the Group became part of the Analytical Division of the Chemical Society. In 1980, following the merger of the Chemical Society, the Royal Institute of Chemistry and the Faraday Society, the Group became part of the Analytical Division of the newly formed Royal Society of Chemistry. The objective of the TMG is to promote awareness of all thermoanalytical, calorimetric and related techniques by a range of activities, including a regular programme of scientific meetings, training workshops and publications.

1.3 TMG Committee Structure

The management of the operation of the Group is the responsibility of the TMG Committee, which is elected at the Annual General Meeting, normally held during the April scientific conference. The Chairman, Vice-Chairman and Immediate Past Chairman of the TMG serve for a period of two years and Committee members for the three years, with the possibility of a further three-year extension. The posts of Secretary and Treasurer are renewable. The Committee hold "face to face" meetings at least twice a year and these are supplemented by telephone conferences where necessary. In addition, sub-committees are formed to deal with specific activities such as training courses and the planning of meetings. The TMG Committee actively welcomes new members, particularly those that can widen its range of thermal analysis and calorimetric expertise.

A complete list of the TMG Chairmen, Secretaries and Treasurers is given in Tables 1.1–1.3, respectively. The Group has been

Table 1.1 A list of TMG Chairs

1965	Robert Mackenzie	Macaulay Institute for Soil Research, Aberdeen
1967	John Redfern	Battersea College of Technology, London
1969	David Dollimore	University of Salford
1971	John Sharp	University of Sheffield
1973	Keith Barrett	ICI Dyestuffs/Patent Office
1975	Fred Wilburn	Pilkington Bros. Ltd.
1977	Dick Still	University of Manchester Institute of Science and Technology
1979	Ted Charsley	Stanton Redcroft Ltd
1981	Graham Clarke	North East Surrey College of Technology
1983	Peter Laye	University of Leeds
1985	Derek Nowell	Hatfield Polytechnic
1987	Peter Haines	Kingston Polytechnic
1989	Jenny Hider	Consultant
1991	Bob Whitehouse	Cabot Plastics
1993	David Morgan	British Geological Survey
1995	Jezz Leckenby	TopoMetrix Corporation
1997	Steve Warrington	Loughborough University
1999	Trevor Lever	TA Instruments
2001	Jim Ford	Liverpool John Moores University
2003	Keyna O'Reilly	University of Oxford
2004	Mark Phipps	TA Instruments
2006	Mike Reading	University of East Anglia
2008	Simon Gaisford	School of Pharmacy, University of London
2010	Ian Priestley	Syngenta Ltd, Huddersfield
2012	Paul Gabbott	PETA Solutions
2014	Vicky Kett	Queen's University Belfast

Table 1.2 A list of TMG Secretaries

Cyril Keattch	1965–1999	Industrial and Laboratory Services
Richard Wilson	1999–2006	SmithKline Beecham
Catherine Barnes	2006–present	GlaxoSmithKline

Table 1.3 A list of TMG Treasurers

Alan Hodgson	1965–1979	Cape Asbestos Company Ltd
Dick Still	1979–1996	University of Manchester Institute of Science and Technology
Mike Richardson	1996–2004	National Physical Laboratory
Simon Gaisford	2004–2006	School of Pharmacy, University of London
Mike O'Neill	2006–2010	University of Bath/Aston University
Simon Gaisford	2010–present	UCL School of Pharmacy

exceptionally fortunate in having long-serving Secretaries and Treasurers who have provided valuable continuity during the 50 years of the Group's history and who have worked tirelessly on the Group's behalf. In particular, mention should be made of the Group's first Secretary, Cyril Keattch, who held the post from 1965 until his untimely death in 1999.

1.4   TMG Meetings

The TMG initially held two or three meetings a year, frequently in conjunction with other groups of the Society of Analytical Chemistry or Royal Institute of Chemistry. The meetings covered a wide range of topics ranging from "Protection of the Environment" to "Controlling Explosives" and were sometimes held in the afternoon or evening. A particular effort was made to meet in different parts of the UK in order to reach as many of the membership as possible. By the 1980's, the Group had developed a pattern of holding a two-day meeting in April and a one-day meeting in November.

In general, the meetings were on specific subject areas and, as a result, TMG members might not attend a meeting for several years if their own field was not featured. In 1995, it was decided that a National Thermal Analysis Conference (TAC) would be held in place of the normal two-day April meeting. This would be open to all areas of thermal analysis and calorimetry and would help promote a regular attendance by Group members. The first TAC was held in Leeds in

Table 1.4 A list of TMG meetings

Meeting	Venue	Organiser
TAC 1996	Leeds Metropolitan University	Ted Charsley
TAC 1997	University of Oxford	Keyna O'Reilly
TAC 1998	University of Surrey	Gary Stevens
TAC 2000	Loughborough University	Duncan Price
TAC 2001	Liverpool John Moores University	Jim Ford
TAC 2002	University of Greenwich	Paul Royall
TAC 2003	University of Huddersfield	Simon Gaisford
TAC 2004	Liverpool John Moores University	Jim Ford
TAC 2005	University of East Anglia	Susan Barker
TAC 2006	School of Pharmacy, University of London	Simon Gaisford
TAC 2007	University of Glasgow	Alan Cooper
TAC 2008	National Physical Laboratory, Teddington	Sam Gnaniah
TAC 2009	University of Bath	Mike O'Neill
TAC 2010	AWE, Aldermaston	Mogon Patel
TAC 2011	Queen's University Belfast	Vicky Kett
TAC 2012	University of Nottingham	Bill MacNaughtan
TAC 2013	University of Greenwich	Milan Antonijevic
TAC 2014	GlaxoSmithKline, Ware	Catherine Barnes
TAC 2015	Churchill College, University of Cambridge	Ian Priestley

1996 and proved a considerable success with over 100 delegates.[2] The pattern was thus established for regular TAC meetings and a list of those organised to date are given in Table 1.4. In addition to the TAC conferences, the Group still holds a one-day meeting in November devoted to a specific topic.

The TMG has also ventured outside the UK on two occasions, holding a joint meeting with the Association Française de Calorimétrie (AFCAT) and the *Groupe de Thermodynamique Expérimentale* in Rennes in 1974 and with the Nordic Society for Thermal Analysis in Bergen in 1986. Both meetings were memorable for the warmth of the welcome that the TMG delegates received and the TMG were delighted to be able to reciprocate by hosting a joint meeting with AFCAT in 1977 at Plymouth Polytechnic. The meeting in Rennes was also notable for a very lively discussion on kinetics, which was in no way hampered by the two main protagonists being unable to speak each other's language.

As will be discussed later, the TMG has also organised four international conferences, two in conjunction with the International

Confederation for Thermal Analysis and Calorimetry (ICTAC) and two with the European Symposium for Thermal Analysis and Calorimetry (ESTAC). In 2005, the TMG became the first national group to hold a meeting with a specialist committee of ICTAC. The topic was sample controlled thermal analysis (SCTA) and the meeting was held at the University of Huddersfield with presentations by the SCTA Committee of ICTAC.

The 50th anniversary of the TMG was marked by a special three-day TAC meeting at Churchill College, Cambridge. This was attended by a large number of Past-Chairmen of the Group, including John Redfern, one of the Group's founders. The occasion was celebrated by a dinner held in the splendid surroundings of Corpus Christi College Dining Hall, where all the Past-Chairmen were presented with Certificates of Appreciation.

This was the fourth anniversary meeting held by the Group. In 1975, the 10th anniversary took place in the Grosvenor Hotel, Chester. This was followed by the 21st anniversary meeting in 1986 at the Bonnington Hotel in London where, at the conference dinner, Stu Bark, the President of the Analytical Division, gave a unique and highly amusing review of all the TMG Chairmen. The 30th anniversary meeting was held at York University, with a memorable dinner in the historic King's Manor. The menu records that the Committee had the foresight to obtain a bar extension until 12.30 am in the halls of residence where the delegates were staying.

1.5 Thermal Analysis Schools/Short Courses

Soon after its formation, the TMG Committee considered the possibility of organizing a residential thermal analysis school. An ideal venue emerged in the form of the newly built Cement and Concrete Research Association in Stoke Poges, Bucks. The first school was held in 1968 and featured lectures, practical sessions and tutorials, which were supplemented by lively evening discussions in the bar. The practical sessions allowed participants to bring their own samples for evaluation. This caused a number of problems, including a smoking furnace having to be removed hastily from the premises by a less than pleased manufacturer and another practical group watching a sample that revealed no transitions at all over a 1000 °C range. It was therefore decided that future TMG schools would provide a structured practical programme and the samples. The practical class to distinguish Stork margarine from butter by both taste and DSC proved to be particularly popular.

The success of the first school resulted in the organization of eight further one-week residential schools over the period 1970 to 1988, many of which were oversubscribed. However, the changing economic climate resulted in the 10th school, due to be held at UMIST in 1991, having to be cancelled. Companies had become unwilling to release their staff for extended periods and it was decided that residential schools would be replaced by one- or two-day workshops, which would include practical sessions.

The last of the workshops was held in 1996, but under the chairmanship of Mike Reading, discussions were held on the possibility of holding one-day lecture courses. The first of these was organised at the University of Glasgow in 2007, immediately before the TAC 2007 Conference and this established the pattern for future short courses, which are now held regularly in association with TAC meetings.

1.6   TMG News and Website

As a means to keep members in touch with the Group's activities, the TMG News was first published in 1981 with Peter Laye as Editor. It was distributed in the form of hard copies until 2002 when, under the editorship of Mogon Patel, it was made available in electronic format. Copies of all the past TMG newsletters are available on the TMG website, http://www.thermalmethodsgroup.org.uk, which is now the main source of information on all the Group's activities.

1.7   TMG Cyril Keattch Award

The TMG Committee began discussions in 1978 on how to help young scientists develop their careers by enabling them to present their work at an international thermal analysis or calorimetry conference. The following year, the TMG Award was established as an essay competition open to scientists under the age of 35 years, working in the UK. The first award was associated with the 2nd European Symposium on Thermal Analysis (ESTA2), which was held in Aberdeen in September 1981, and the winner was Eddie Paterson from the Macaulay Institute for Soil Research, Aberdeen.

The award, which was renamed the Cyril Keattch Award in 2002 in memory of the Group's long serving first Secretary, has been regularly associated with both ESTAC and ICTAC meetings and to date has supported a total of 17 scientists at conferences as far afield as Japan and Brazil. The list of winners is shown in Table 1.5 and the wide range

Table 1.5   A list of TMG Award winners

1981	ESTA 2, Scotland	Eddie Paterson	Macaulay Institute for Soil Research

The Value of DSC in Assessing the Physical and Chemical Properties of Particle Surfaces.

1982	7th ICTA, Germany	Tom Taylor	University of Salford

Studies on the Degradation of Nickel Nitrate Hexahydrate using Thermal Methods.

1984	ESTAC 3, Switzerland	Atiq Rahman	University of Aberdeen

Application of Thermal Methods in Surface Chemical Investigations of Zirconium Gels.

1987	ESTAC 4, GDR	Ed Gimzewski	BP Research Centre

Thermal Analysis in Reactive Atmospheres.

1988	9th ICTA, Israel	Mike Reading	ICI Paints

The Kinetics of Heterogeneous Solid-state Decomposition Reactions: A New Way Forward?

1992	10th ICTA, UK	Tony Ryan	UMIST

Simultaneous Small-Angle X-Ray Scattering and Wide-Angle X-ray Diffraction.

1994	ESTAC 6, Italy	Michael Ewell	UMIST

Forced-Adiabatic Sampling Environments: Useful Tools for the Study of Structure Development during Polymerisation.

1996	11th ICTAC, USA	Gary Foster	Birkbeck College

Simultaneous Non-Invasive Microwave Dielectric Spectroscopy and Dynamic Mechanical Analysis for Studying Drying Processes in Complex Heterogeneous Materials.

1998	ESTAC 7, Hungary	Chris Allen	University of Oxford

Calorimetric Control of Aluminium Casting Quality.

2000	12th ICTAC, Denmark	Zhong Jiang	University of Aberdeen

Temperature Modulated Differential Scanning Calorimetry: Modelling the Effects of Heat Transfer on TMDSC Measurements.

2002	ESTAC, Spain	Vicky Kett	Queen's University Belfast

The Application of Thermoanalytical Techniques to the Study of the Freeze Drying Process.

2004	13th ICTAC 13, Sardinia	Laura Waters	University of Huddersfield

Saturation Determination of Micellar Systems Using Isothermal Titration Calorimetry.

Table 1.5 *(Continued)*

2006	ESTAC 9, Poland	Mike O'Neill	School of Pharmacy, University of London

Is complexity an issue? The quantitative analysis of calorimetric data.

2008	14th ICTAC, Brazil	Louise Grisedale	University of East Anglia

Photothermal Microspectroscopy (PTMS); A New Technique for Spatially Differentiating Between Crystalline and Amorphous Materials.

2010	ESTAC 10, the Netherlands	Paul Nevitt	AWE Harwell

Applications of Thermal Analysis and Calorimetry to the Study of Metal Hydrides, Deuterides And Tritides.

2012	15th ICTAC, Japan	Fuad Hajii	University of Nottingham

The Intrinsic Influence of *N*-Methylmorpholine-*N*-Oxide on the Phase Transitions of Native and Physically Modified Starch.

2014	ESTAC 11, Finland	Katie Hardie	University of Manchester

DSC as a Primary Tool for the Development of a Semi-Permanent Hair Straightening Technology.

of topics presented provides a good illustration of the wide scope of thermal analysis and calorimetric techniques. It is gratifying to note that two of the winners have gone on to become TMG Chairmen.

The Cyril Keattch Award competition is currently held biennially. It consists of an award certificate and a specified sum of money towards the conference registration fee, travel and living expenses at the conference. The current regulations state that candidates for the award must be resident in the UK and should have normally worked for 10 years or less at the post-graduate level. The winner is chosen by inviting the authors of the best three written submissions to give their presentations at the TAC Conference before the nominated international meeting. The next award will be associated with the 16th ICTAC Congress to be held in Orlando, Florida, USA, in August 2016.

1.8 Thermal Analysis and Calorimetry Equipment Manufacturers

From the very earliest days, the TMG has been very fortunate to have been enthusiastically supported by the instrument manufacturers.

It would not have been possible to hold the thermal analysis schools without their equipment and skilled technical support and the instrument exhibition remains an integral part of the TAC conferences.

In addition, instrument company staff make a valuable scientific contribution to both the TMG meetings and to the work of the Committee with an instrument manufacturers' representative serving on the Committee. Their importance to the Group may be judged from the fact that five TMG Chairmen have been from instrument companies. Workers in the field also benefit from the exceptionally large range of technical information and application sheets that are provided by the instrument companies.

1.9   International Confederation for Thermal Analysis and Calorimetry

The TMG has been closely associated with the International Confederation for Thermal Analysis and Calorimetry (ICTAC) since its inception. After correspondence with thermal analysts in thirty countries, the unanimous support for an international thermal analysis meeting led Robert Mackenzie and John Redfern to decide to organise the First International Conference on Thermal Analysis in Aberdeen in 1965.[3] With Robert as Chairman and John Redfern as Secretary, they were joined on the organising committee by Rudolf Bárta (Czechoslovakia), Leo Berg (USSR), Lázsló Erdey (Hungary), Connie Murphy (USA) and Toshio Sudo (Japan), with Bruce Mitchell providing invaluable local support.

The conference, which was held in the Natural Philosophy Building of the University of Aberdeen, was attended by nearly 300 scientists from 29 countries. The considerable success of the meeting resulted in a Second International Conference being held in Worcester, Mass, USA, in 1968 where the International Confederation for Thermal Analysis and Calorimetry (ICTA) formally came into being.

In 1992, ICTA returned to the UK, and the TMG hosted the 10th ICTA Congress at the University of Hertfordshire, with Derek Nowell as Organising Chairman. At this conference, which attracted over 250 delegates from 23 countries, ICTA formally changed its name to ICTAC to reflect the large calorimetric component of its membership.

The TMG is represented on the ICTAC Council by an Affiliate Councillor who can serve for up to two four-year terms. In addition, a number of official posts within ICTAC have been held by TMG members including President (Ted Charsley), Treasurer (Robert

Mackenzie, Dick Still, John Crighton), Nomenclature Committee Chairman (Robert Mackenzie, John Sharp, Ed Gimzewski, Trevor Lever), Standardisation Committee Chairman (Ted Charsley, Mike Richardson), Awards Committee Chairman (David Morgan) and ICTAC Newsletter Editor (Robert Mackenzie, Cyril Keattch, Stuart Du Kamp). Robert Mackenzie and Ted Charsley were awarded Honorary Lifetime Membership of ICTAC. The unique contribution of Robert Mackenzie to both thermal analysis and ICTAC was recognised by the establishment of the Robert Mackenzie Memorial Lecture in 2000, which is given after the opening ceremony at each ICTAC Congress.

1.10   The European Symposium on Thermal Analysis and Calorimetry

The European Symposium on Thermal Analysis and Calorimetry owes its origins to a decision by the TMG to organise an international meeting on thermal analysis in the UK and the first European Symposium on Thermal Analysis (ESTA 1) was held at Salford University in 1976.[4] David Dollimore was the Organising Chairman and his connections with the Boddingtons Brewery ensured a memorable conference dinner. An open meeting held during the conference enthusiastically agreed that further symposia should be organised and the Thermal Analysis Society of the Soviet Union offered to host the next conference. However, by 1979, it became clear that they would not be able to undertake the task and after the TMG Chairman Dick Still had contacted all the European thermal analysis and calorimetry groups, the TMG decided to organise the 2nd ESTA. This was held at Aberdeen University in 1981 with Fred Glasser as the Organising Chairman.

The future of ESTA was, however, still in doubt but at ESTA 2, a meeting of representatives from all the European thermal analysis societies present was organised and the ESTA Committee was formed. There was a very positive response to continuing to organise future meetings and invitations to host 3rd and 4th ESTA's in Interlaken and Jena, respectively, were presented at the meeting by Switzerland and the German Democratic Republic. The future of ESTA was thus ensured and eleven symposia have been organised to date, with calorimetry incorporated into the title in 1982.

A small part of the UK is always present at an ESTAC conference in the form of the ESTAC Stones. These are pebbles gathered on the

Aberdeen seashore during the ESTA 2 meeting by Hans Oswald and Erwin Marti from Switzerland who had the idea that they could become the symbol of the symposium. They are mounted in an acrylic container and are presented by the ESTAC Chairman to his or her successor at the closing dinner.

1.11   High Alumina Cement Crisis

In 1974, many TMG members became involved in the largest thermal analysis test programme ever undertaken. This followed the collapse of the roofs of two school building constructed with high alumina cement (HAC) concrete beams. Harry Midgley at the Building Research Establishment, who also supervised the implementation of the test programme, developed a test method using DTA to measure the degree of conversion of samples of HAC, taken *in situ* from beams. The programme called for tens of thousands of samples to be tested as rapidly as possible and was later extended from DTA to include both DSC and DTG as test methods.

As an increasing number of laboratories enrolled in the programme, it became clear to the TMG that a detailed set of guidelines was required covering the test methods and interpretation and reporting of results. A working party was set up in November 1974 and in April 1975, with the approval of the Chemical Society, published "Recommendations for the Reporting of High Alumina Cement Concrete Samples by Thermoanalytical Techniques".[5]

1.12   The Future

We hope that this brief review has shown that the TMG has fully realised its original idea of providing a UK forum for thermal analysts and calorimetrists. The Group has adapted successfully to the many changes that have taken place and now looks forward with optimism to the next 50 years and the challenges ahead. One of the aims of the TMG is to broaden further its membership and if you are not yet a member, we would warmly invite you to join the Group. We are sure that you will find that this will be both professionally and personally rewarding and you will be able to help shape the next phase of the TMG's history. Details of how to join and of all the Group's activities can be found on the TMG website at http://www.thermalmethodsgroup.org.uk

Acknowledgements

We would like to thank Professor Peter Laye of the University of Huddersfield for his helpful comments.

References

1. C. J. Keattch, *Anal. Proc.*, 1988, **25**, 342.
2. *Current Research & Development in Thermal Analysis and Calorimetry*, ed. E. L. Charsley and P. G. Laye, Proceedings 1st TAC Conference, Thermal Methods Group, Leeds, 1996, *Thermochim. Acta*, 1977, **294**, 1–131.
3. R. C. Mackenzie, *J. Therm. Anal.*, 1993, **40**, 5.
4. E. L. Charsley, *J. Therm. Anal. Calorim.*, 2011, **105**, 727.
5. F. W. Wilburn, C. J. Keattch, H. G. Midgley and E. L. Charsley, Thermal Methods Group, Analytical Division, Chemical Society, Special Publication, 1975.

2 Thermal Analysis Nomenclature

Trevor Lever

Trevor Lever Consulting (TLC), 1 Hope Close, Wells,
Somerset, BA5 2FH, UK
Email: trevor@trevorleverconsulting.com

2.1 Introduction

The Oxford English Dictionary defines nomenclature as "The devising or choosing of names for things, especially in science". This definition presents two approaches to us: one where we devise a systematic and logical approach and the other where someone, or somebody, chooses what "things" should be called.

Getting a global agreement on nomenclature therefore presents a challenge. As ideas and techniques develop, new language is required to describe and define, so those who develop new "things" get the opportunity to name it. That seems fair, doesn't it? However, as scientists, we desire a systematic approach in the way we work. This gives us some structure and familiarity. This also seems reasonable as opposed to a random naming of 'things' based on whim or personal preference.

So there are these two opposing strands at work in any developing field: the structured approach *versus* the personal approach. Who wins? Who gets to have their name for a new technique or experimental approach recognised in the official nomenclature?

In the early years of thermal analysis, when no commercial equipment was readily available, scientists built their own equipment

Principles of Thermal Analysis and Calorimetry: 2nd Edition
Edited by Simon Gaisford, Vicky Kett and Peter Haines
© The Royal Society of Chemistry 2016
Published by the Royal Society of Chemistry, www.rsc.org

to support their research. These "lab-built" instruments allowed the pioneers of the technique to make measurements and publish, but often led to problems with others repeating their work, as each instrument was different in construction and operation. This was recognised by discussions within the Executive Committee of the International Confederation for Thermal Analysis (ICTA, now the International Confederation for Thermal Analysis and Calorimetry, ICTAC), which set up a sub-committee to deal with nomenclature in conjunction with the Thermal Methods Group (TMG). This new committee, chaired by Robert C. Mackenzie with Cyril J. Keattch as secretary, became a sub-committee of the TMG and was supported by an international group of corresponding members.[1] The role of Robert Mackenzie cannot be over-estimated here in steering the early discussions on nomenclature, and his advice is acknowledged and included in the official ICTAC-IUPAC (International Union for Pure and Applied Chemistry) approved 2014 nomenclature guidelines.

Today, ICTAC continues to have a nomenclature committee whose role it is to discuss and agree the names for things and present their proposals back to the scientific community. Ultimately, it is the community at large who will decide if the work of the nomenclature committee is successful, by adopting and using the proposals in the reporting of their work and research. Understandably, getting agreement requires compromise. Over the tenure of the nomenclature committee, the discussion ranges over science, linguistics and even philosophy in an attempt to reach a consensus.

In 1969, the official ICTA[2]/IUPAC[3] definition of thermal analysis was: *Thermal Analysis: a term covering a group of techniques in which a physical property of a substance and/or its reaction product(s) is measured as a function of temperature whilst the substance is subjected to a controlled temperature programme.*

Today, we would see a number of problems with this definition; the two main issues being that (a) isothermal measurements would be excluded from the definition of thermal analysis, as well as (b) any measurement that is "sample-controlled" rather than "temperature-controlled" is not covered.

In 1991, a new definition was proposed (although never officially endorsed by the ICTAC council) in the ICTAC booklet "For Better Thermal Analysis":[4] *Thermal Analysis: a group of techniques in which a property of the sample is monitored against time or temperature while the temperature of the sample, in a specified atmosphere, is programmed. The programme may involve heating or cooling at a fixed rate of temperature change, or holding the temperature constant, or any sequence of these.*

This definition reflects the change in complexity of experiments possible with the use of computers and micro-processors to control the temperature programme and it also embraces isothermal measurements as part of thermal analysis.

One of the reasons the definition was never fully adopted was the debate around including isothermal measurements into the definition. Few thermal analysts saw an issue with this, but many scientists in other analytical sciences saw the definition as including room temperature experiments and so embracing much of spectroscopy and chromatography! Although this was not the intent in the definition, it was an unfortunate consequence. Also, the definition still excluded sample-controlled experiments, which were gaining importance and were increasingly reported in the literature.

So, a new ICTAC nomenclature committee was instigated in 2001 and set the task of rationalising all of the previous committee's work and delivering a definition and document that would cover current thermal analysis practice.

A significant part of the early work of this new committee was to focus on the definition of thermal analysis. There was a strong desire to simplify the definition, to cover dynamic and isothermal experiments and to include sample-controlled measurements as well. There are many pages of suggestions, counter suggestions and concerns around potential definitions in the minutes of committee meetings from that time. A key breakthrough was the introduction of the word "relationship" within the definition, specifically "the relationship between a sample property and its temperature" as this gave equal emphasis to temperature-controlled and sample-controlled experiments. The output of most thermal analysis experiments is a plot of a property against temperature and the curve describes the "relationship" within the definition.

For a while, the proposed definition also included isothermal experiments: "…as the sample is heated, held at constant temperature, or cooled in a controlled manner". Again, the same concerns were raised against this definition as the one proposed back in 1991, and so in the end "held at constant temperature" was removed.

After considerable effort and debate, the final definition was reached, and as with much work in the field of nomenclature, it is a reasonable compromise: *Thermal Analysis: is the study of the relationship between a sample property and its temperature as the sample is heated or cooled in a controlled manner.*

Not only does this definition exclude any isothermal experiments, it also excludes any reference to experimental parameters or atmosphere control. This is intentional, and although it is desirable to keep all experimental variables to a minimum, this does not need to be mentioned in the definition. The definition also has the benefit of being flexible for developing techniques and is concise.

In 2006, the ICTAC Council adopted this definition and the full nomenclature proposals of the committee. This was published, along with an initial outline for the nomenclature of calorimetry, in 2008.[5]

Once approved by the ICTAC Council, Jean Rouquerol (ICTAC Past President) oversaw the adoption of the new nomenclature within the IUPAC. There were some (small) changes and questions to answer and defend, and this process also took several years moving through the IUPAC ratification process.

In April 2014, following IUPAC adoption, the full nomenclature was published in Pure and Applied Chemistry.[6]

The Thermal Methods Group of the UK can be rightly proud of its contribution to the nomenclature process and the work in leading and supporting the other thermal analysis groups and representatives from around the globe in producing this document and reaching this consensus.

References

1. R. C. Mackenzie, Origin and Development of ICTA, *J. Therm. Anal.*, 1993, **40**, 9.
2. R. C. Mackenzie, *Talanta*, 1969, **16**, 1227.
3. H. W. Thompson, *Pure Appl. Chem.*, 1974, **37**, 439.
4. J. O. Hill, 1991, *For Better Thermal Analysis and Calorimetry*, ICTAC, 3rd edn.
5. *Handbook of Thermal Analysis and Calorimetry*, ed. M. E. Brown and P. K. Gallagher, Elsevier, 2008, vol. 5, ch. 2.
6. T. Lever, P. Haines, J. Rouquerol, E. Charsley, P. Van Eckeren and D. Burlett, *Pure Appl. Chem.*, 2014, **86**, 545.

3 Thermogravimetry

Vicky L. Kett[a] and Duncan M. Price*[b]

[a] School of Pharmacy, Queen's University Belfast, 97 Lisburn Road, Belfast BT9 7BL, UK; [b] Edwards, Kenn Road, Clevedon BS21 6TH, UK
*Email: duncan.price@edwardsvacuum.com

3.1 Introduction

Mass and heat are some of the oldest concepts known to humanity. The discovery of fire provided ample evidence of the effect of heat on materials and in particular mass changes brought about by burning.[1] The careful study of such processes led to the abandonment of the concept of phlogiston and ultimately evolution of chemistry out of alchemy. On a more mundane level, classical gravimetric analysis has a long pedigree for determining the constitution and composition of substances by dissolution and precipitation of components which are assayed after removal of solvents by drying. Thermogravimetry simply exploits the use of heat to bring about mass changes by chemical decomposition or reaction with the atmosphere and thus tells the investigator something useful about the sample under investigation.

Thermogravimetry or thermogravimetric analysis is one of the most popular thermal methods. The terms are used interchangeably and often abbreviated to TG or TGA (although the former can easily be confused with the glass transition temperature "T_g"). The origins of thermogravimetry date back to 1912 when Urbain described the first equipment that could continuously heat and weigh a sample under a controlled gas atmosphere.[2] Three years later, Honda coined the term

Principles of Thermal Analysis and Calorimetry: 2nd Edition
Edited by Simon Gaisford, Vicky Kett and Peter Haines
© The Royal Society of Chemistry 2016
Published by the Royal Society of Chemistry, www.rsc.org

"thermobalance" for his instrument which was also capable of reducing the heating rate when a mass change was occurring leading to a form of sample-controlled thermal analysis.[3] The technique was rapidly adopted for inorganic chemical analysis, reaching its zenith in Duval's monumental description of over 1000 gravimetric precipitates of nearly 80 elements.[4] Applications to polymers and pharmaceuticals blossomed after the 2nd World War and became more widespread with the introduction of commercial instrumentation beginning in the 1960s.[5]

3.2 Instrumentation

A schematic diagram of a thermobalance is shown in Figure 3.1. The device consists of a sensitive recording balance with provision for the sample to be heated in a gas-tight oven or furnace that is purged with an atmosphere of inert or reactive gas such as nitrogen or air. There will be some electronics for temperature control and signal processing. A computer will be used as a data station for the display and archiving of results; this can typically be interfaced to several thermal analysis instruments and allows the operator to define experimental parameters such as the temperature profile and gas composition.

3.2.1 Balance Mechanism

Null-point balances are the most common type of mechanism in most instruments. These employ an electromagnetic compensation principle whereby a beam carrying the sample and countermass is suspended from the coil of a galvanometer.[6] As the mass of the sample changes, the beam turns on a pivot causing a variation in the current from a

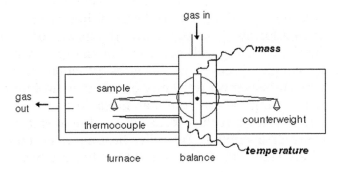

Figure 3.1 Schematic diagram of a thermobalance showing arrangement of purge gas flow and relationship between sample and thermocouple.

photodetector–shutter–lamp arrangement. This is used in a servoloop to apply a restoring force through to the coil. The change in current (or voltage) required to maintain equilibrium is proportional to the mass change in the sample. This analogue signal is amplified and digitised for further processing. Typical sample sizes range from between 1 and 100 mg, although obtaining a truly representative sample of the material under investigation may be difficult as the amount analysed decreases. Specialised thermobalances designed to accommodate larger mass ranges have been described[7,8] as have devices with a magnetic suspension arrangement that isolate the sample holder from the balance mechanism.[9] Other specialised balance mechanisms based on strain gauges[10] or quartz springs[11] have been used for specialised applications. Calibration of the balance mechanism is infrequently required but is easily checked using standard milligram masses often supplied with the instrument for this purpose.[12] The sensitivity of most balances is of the order of 1 microgram and linearity across the full mass range is critical since results are often quoted as mass fraction normalised to the initial starting mass; it is also important to tare the balance against the empty sample holder before measurement. Many thermobalances can be fitted with automatic sample changers, thus permitting unattended operation and increased productivity. For hygroscopic samples, it is usual to encapsulate the sample, which is then pierced immediately before loading onto the balance.

3.2.2 Furnace

Whereas most instruments use a balance mechanism of a standard design, the furnace is often designed for a specific temperature range or response time. Sub-ambient operation is not usually possible unless the instrument is equipped for simultaneous thermogravimetry-differential scanning calorimetry (TG-DSC), although many furnaces have the facility for air or water cooling (by means of a jacket) so that they may be cooled quickly at the end of an experiment in order to increase sample throughput. Furnaces generally employ non-inductively wound electrical resistance elements to avoid interaction with magnetic samples. Low thermal mass furnaces allow rapid changes in temperature at the expense of good temperature stability under isothermal operation. Typically, heating rates range up to $50\,°C\,min^{-1}$ are employed although infra-red heating has been used to achieve a more rapid response.[13]

Several different arrangements of the balance and furnace are possible (Figure 3.2). The sample may hang down from the balance

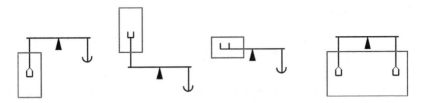

Figure 3.2 Different furnace configurations (left–right): below balance, above balance, horizontal (*c.f.* Figure 3.1) and symmetrical.

into the furnace, or a top loading design may be used. Horizontal configurations may also be used in order to reduce the gas flow affecting the apparent mass of the sample during heating by impinging on the sample and thus acting as a piston. Alternatively, the inlet and outlet lines for the purge gas may be arranged to direct the flow of gas across the sample rather than up or down a vertical furnace. The density of the gas decreases with increasing temperature, this reduces the upthrust on the sample (Archimedes' Principle) leading to an apparent increase in mass. The gas flow path may also change with temperature. Using small samples (<10 mg) or performing a blank measurement with a similar volume of inert material can reduce these effects. An alternative approach is to employ a symmetrical design whereby the sample holder and countermass both hang down into the same or separate furnaces.[14]

3.2.3 Atmosphere Control

Provision must be made for some means of purging the furnace with a controlled atmosphere. Operation under an inert atmosphere (such as nitrogen) may not always be possible since the furnace may not be completely air tight and may also retain trapped oxygen. Even high purity nitrogen may not be completely oxygen-free. This may be of concern for some materials that are sensitive to oxidation.[15] Helium can be used for good heat transfer between the furnace and sample, which might be a consideration for experiments that involve fast or rapidly-changing heating rates. Gas flow rates are controlled by rotameters or mass flow controllers, and there is often some provision for switching between gases for compositional investigations. Purge gas flow rates should be sufficient to sweep volatiles from the furnace, but not so excessive as to cool the sample. High pressure and vacuum thermobalances have been described,[16–18] these have applications for the study of industrial processes that occur under elevated or reduced pressures.[19,20]

3.2.4 Sample Holder

Samples are usually contained within small crucibles made of refractory materials such as quartz, alumina or platinum, which are chemically unreactive towards most materials. Caution must be exercised with platinum crucibles as many molten metals can alloy with the container and carbonaceous residues will eventually attack platinum at high temperature particularly if the crucibles are heated in a Bunsen burner flame to clean them.[21] Platinum can also have a catalytic effect on some chemical reactions. Less expensive, disposable aluminium holders may be used for organic materials that usually decompose below the melting temperature of aluminium (660 °C). In all cases, it is important to allow good interaction between the sample and purge gas so that the build-up of decomposition products in the immediate vicinity of the sample is avoided.

3.2.5 Temperature Measurement and Calibration

The temperature of the sample is usually measured by a thermocouple placed as close as possible to the sample without interfering with the operation of the balance. It is unreliable to use the furnace temperature as a measure of sample temperature as this will inevitably be in advance of the sample's true temperature because of heat transfer considerations. In the event of a chemical reaction occurring, the sample temperature will be affected by the enthalpy of reaction and may be above the furnace temperature for exothermic decompositions. Temperature calibration should be carried out regularly, according to a recommended procedure,[22] and always under the same conditions (*e.g.* heating rate, purge gas, thermocouple location *etc.*) as those used for measurements. Achieving reproducibility of temperature is often more important than absolute accuracy, particularly for kinetic studies which may use data obtained from several experiments.

One common method for temperature calibration is to determine the Curie temperature of a metal or other magnetic substance by heating it whilst under the influence of an external magnetic field.[23] An apparent change in weight will occur due to the attraction between them which will disappear when the material passes from the ferromagnetic state to the paramagnetic state (Figure 3.3).

Other methods of temperature calibration employ the melting of a material of known melting point. Thermogravimetry cannot determine melting temperatures directly, but if a fusible link is made from, for example, indium wire and hung from the sample arm in such a

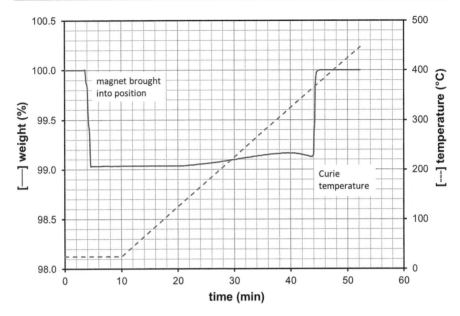

Figure 3.3 Temperature calibration using the Curie temperature of nickel.

manner that it will fall off when the metal melts, this will cause a disturbance in the mass measured.[21] Pseudo-calorimetric methods can also be employed if the sample thermocouple is close enough to the specimen such that the recorded temperature profile is disturbed by the enthalpy of fusion of the specimen (Figure 3.4).[24] Some instruments exploit the deviation from linearity of heating caused by enthalpic effects within the sample detected by a close-coupled sample temperature sensor to provide an additional signal akin to differential thermal analysis (DTA). This treatment is shown in Figure 3.5 for the data presented in Figure 3.4. True simultaneous TG-DTA and TG-DSC instruments employ dedicated arrangements for measuring differential temperature or heat flow. These can adversely affect the sensitivity of the balance although Calvet-type volumetric heat flux sensors do not require lead-out wires from the balance suspension and are less invasive.[25]

3.2.6 Ancillary Equipment

Modern thermobalances employ a computer for data acquisition, although many also have built-in status displays that allow a limited amount of stand-alone operations either *via* touch screens or other controls. The computer acts as a data station so that results may be

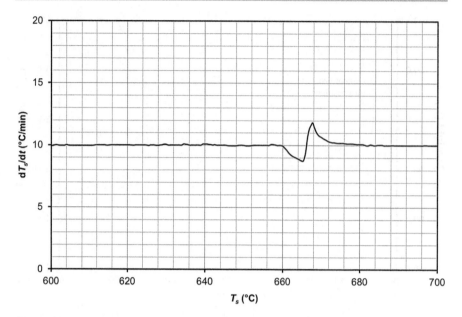

Figure 3.4   Pseudo-calorimetric temperature calibration with aluminium. T_s is the sample temperature.

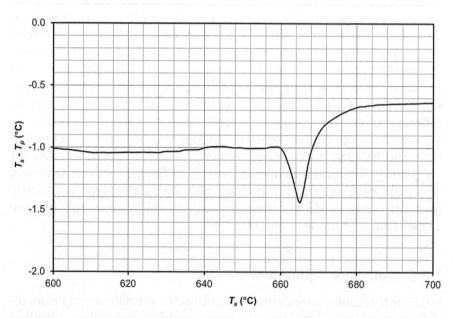

Figure 3.5   "Single differential thermal analysis" curve generated from subtracting the programme temperature (T_p) from the sample temperature (T_s) in Figure 3.4.

stored for future reference and subsequent treatment. It is usual to be able to plot the sample mass (directly or as a percentage of its initial mass) as a function of time or temperature. Calculation and display of the first derivative of sample mass (m) as a function of time or temperature (dm/dt or dm/dT) is a useful facility in identifying regions of interest. It is common to plot $-dm/dt$ or $-dm/dT$ so that maxima in rate of mass loss appear as positive peaks. At a basic level, characteristic temperatures and magnitudes of mass changes are reported. More sophisticated treatments extend to kinetic modelling of processes although it is essential to have a thorough understanding of any mathematical manipulation of data before any reliance is placed on the results.

3.3 Basic Experiments

A key consideration in all thermoanalytical measurements is the importance of reporting all results with sufficient detail including sample history, measurement conditions, equipment and data treatment, so as to enable their repetition and, if necessary, reconcile any observed differences in outcome with changes in operational procedure. Careful attention to consistency in experimental procedures usually results in good repeatability. Conversely, studying the effect of deliberate alterations in such factors as the heating rate or sample atmosphere can give valuable insights into the nature of the observed reactions.

Data from the thermogravimetric analysis of copper(II) sulfate pentahydrate, $CuSO_4 \cdot 5H_2O$, readily illustrate the salient features of the technique and provide a starting point for the discussion of instrumental and sample-related effects. Figure 3.6 shows a plot of sample mass (as a percentage of the initial mass) as a function of temperature. The derivative mass change curve ($-dm/dT$) indicates that five mass loss processes occur during the experiment, these being progressive dehydration (steps 1–3 in Scheme 3.1) followed by decomposition of the anhydrous salt to copper(II) oxide and then reduction to form copper(I) oxide.

The effect of changing the heating rate can be considered in terms of the overlapping sequential mass losses for the dehydration processes. These would be better resolved by heating the sample more slowly (this is the philosophy behind sample-controlled thermogravimetry, discussed later). A higher purge gas flow rate might remove any volatiles more efficiently, while switching from an inert to

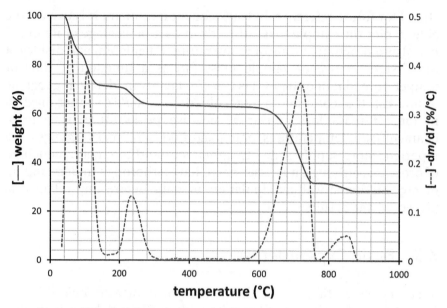

Figure 3.6   Plot of mass and rate of mass loss ($-dm/dT$) *vs.* temperature for copper sulfate pentahydrate for a heating rate of 5 °C min^{-1} under nitrogen.

$$CuSO_4 \cdot 5H_2O \rightarrow CuSO_4 \cdot 3H_2O + 2H_2O \qquad \text{(step 1, 40–80 °C)}$$

$$CuSO_4 \cdot 3H_2O \rightarrow CuSO_4 \cdot H_2O + 2H_2O \qquad \text{(step 2, 80–140 °C)}$$

$$CuSO_4 \cdot H_2O \rightarrow CuSO_4 + H_2O \qquad \text{(step 3, 180–300 °C)}$$

$$CuSO_4 \rightarrow CuO + SO_2 + \tfrac{1}{2}O_2 \qquad \text{(step 4, 560–770 °C)}$$

$$CuO \rightarrow Cu_2O + \tfrac{1}{2}O_2 \qquad \text{(step 5, 770–880 °C)}$$

Scheme 3.1   Thermal decomposition of copper(II) sulfate pentahydrate.

an oxidising atmosphere would have significant impact on the final mass change. A tightly-confining crucible would also affect reactions by maintaining close interactions between the sample and its decomposition products (a "self-generated" atmosphere).[26]

The sample itself can introduce a number of factors that can influence the results of measurement. Whilst it might be attractive to increase the resolution of small mass changes by using a larger sample, this can cause smearing of overlapping processes because of the difficulty associated with heating a large sample uniformly. Furthermore, there may be poor exchange of gaseous decomposition products with the furnace atmosphere. Coarsely ground powders can suffer from the same effect, although finely grinding a specimen may lead to unwanted changes in morphology or composition (*e.g.* loss of moisture). For these reasons, any form of pre-treatment should be described.

Some materials may show ferromagnetic behaviour, which will disappear at their Curie temperature (alternatively, some materials may be transformed into ferromagnetic phases on reaction[27])—such systems can be studied by the technique known as "thermo-magnetometry."[28,29] The use of an external magnetic field to calibrate the temperature scale of a thermobalance was described in Section 3.2.5 and there are numerous applications of this method for the characterisation of magnetic materials.[30–32]

3.3.1 Thermal Stability Assessment

One of the most valuable uses of thermogravimetry is illustrated by the results obtained from a series of ammonium salts (ammonium hydrogen fluoride, ammonium fluoride and ammonium fluoro-silicate) shown in Figure 3.7. These materials are frequently encountered in the exhaust from pumps used to evacuate semi-conductor processing chambers and arise from the interaction of fluorine-containing etch gases and ammonia-containing deposition gases used in silicon nitride production. Thermogravimetry affords a

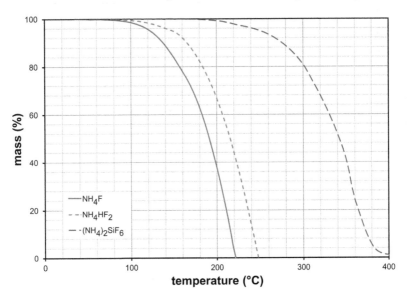

Figure 3.7 Plots of mass *vs.* temperature for ammonium fluoride, ammonium hydrogen fluoride and ammonium hexafluorosilicate, 10 °C min^{-1} in nitrogen.

useful guide to the temperatures required to keep these substances from condensing and causing pipe blockages, although one can also imagine comparing the mass loss curve of a deposit found in a pipe with characteristic curves of candidate materials to identify the unknown substance. This illustrates the concept of using thermogravimetry to "fingerprint" materials and is particularly useful when the derivative mass loss curve is used to identify minor processes arising from subtle differences in compostion.[33]

3.3.2 Compositional Analysis

Certain classes of materials lend themselves to compositional analysis by thermogravimetry.[34] These include filled organic polymers whereby the matrix can be completely decomposed by heating to leave behind only inorganic filler or reinforcement, such as talc, glass fibre, *etc.* Although these determinations could readily be performed in a muffle furnace, using a thermobalance to obtain the mass loss profile can yield additional information about the matrix. An example of this is shown in Figure 3.8 for a series of poly(ethylene-*co*-vinyl acetate) copolymers. The vinyl acetate component shows a characteristic mass loss associated with elimination of acetic acid before the main chain

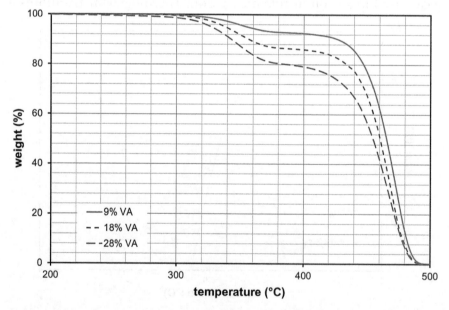

Figure 3.8 Plots of mass *vs.* temperature for a series of poly(ethylene-*co*-vinyl acetate) copolymers of different vinyl acetate (VA) content, 5 °C min^{-1} in nitrogen.

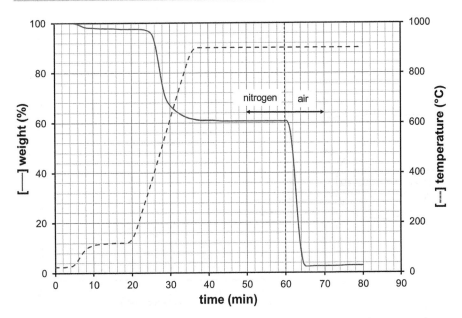

Figure 3.9 Proximate analysis of wood charcoal. The sample atmosphere is changed from nitrogen to air 60 minutes into the experiment.

degradation occurs. In this case, thermogravimetry can be used as a quantitative means of estimating copolymer composition.

Thermogravimetry comes into its own when more elaborate temperature programmes and gas switching are employed. Such an experiment is illustrated in Figure 3.9 for the proximate analysis of charcoal. Here, the sample is heated to 120 °C and held at this temperature for 10 minutes in order to measure its moisture content. The temperature is then ramped to 900 °C to drive off volatile organic material. Finally the purge gas is switched from nitrogen to air so that the carbonaceous residue is burned off and only an inorganic ash residue remains. Figure 3.10 shows the results of a series of analyses taken from a 50 mm diameter piece of wood that had been heated for five hours.[35] This example demonstrates how thermogravimetry can be used to examine small samples taken from a larger specimen in order to detect compositional variations within the bulk. A similar approach is used to classify coal according to its moisture, volatiles, carbon and ash content.[36,37]

3.3.3 Lifetime Prediction

One of the most widespread uses of thermogravimetry beyond that of compositional analysis and thermal stability assessment is to predict

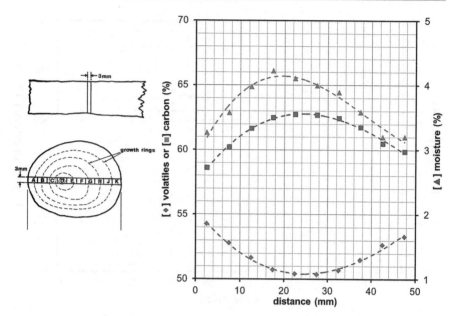

Figure 3.10 Composition of charcoal samples taken from a section through a 50 mm diameter piece of carbonised wood.

the lifetime of materials. The simplest approach is to perform iso-thermal measurements at elevated temperatures and measure the time taken for a certain extent of mass loss to occur. Several experiments may be carried out at different temperatures so as to obtain a table of lifetime *versus* temperature. Whilst such data are readily obtained, they rarely span the temperature range of interest (indeed to do so may require inordinately long experiments at lower temperatures) and therefore some means of extrapolation is required.

The temperature dependence of chemical processes may be readily expressed in terms of the Arrhenius equation:[38]

$$k = A \exp(-E_a/RT) \qquad (3.1)$$

where k is the rate constant, R the gas constant and T the thermodynamic temperature (K). Values of the Arrhenius parameters (E_a and A) provide measures of the magnitude of the energy barrier to reaction (the activation energy, E_a) and the frequency of the occurrence of a condition that may lead to reaction (the frequency factor, A). The rate constant k is defined by the relationship between the rate of reaction ($d\alpha/dt$) and the extent of conversion or fraction reacted (α). Decompositions studied by thermogravimetry are generally hetero-geneous reactions since the sample is solid (or molten) and the products are gases. A general relation to describe the relationship between $d\alpha/dt$ and α has been described by Šesták and Berggren:[39]

$$d\alpha/dt = k\,\alpha^m(1-\alpha)^n[-\ln(1-\alpha)]^p \tag{3.2}$$

Starting from this basic form, it is possible to derive various subclasses of rate equation such as first order decay, nucleation and growth, *etc.* by changing the values of m, n and p.[40] Isothermal experiments provide the means to determine the form of the kinetic equation though discrimination between different models is not straightforward.[41] A simpler approach is to substitute the reciprocal of the isothermal lifetime for the rate constant in the equation and extrapolate the data to the region of interest. Caution should be exercised when interpreting such data, especially when extrapolating to lower temperatures where the degradation mechanism may have changed. This is especially important if the original measurements were made above a phase transition *e.g.* the material's melting point.[42]

Isothermal measurements suffer from the drawback of being time-consuming. There are also difficulties in bringing the sample and apparatus to the required temperature without some decomposition of the sample having already taken place. A more convenient approach to studying degradation kinetics is to employ data from conventional linear rising temperature thermogravimetry. Many such methods have been proposed, but the most popular strategy is that described by Ozawa[43] and Flynn and Wall,[44] which has been incorporated into a standard method.[45,46] Essentially, separate measurements are carried out at different linear heating rates and the temperatures at which a set percentage mass loss occurs noted (Figure 3.11).

These are then plotted as a function of heating rate (dT/dt) and the E_a determined by an iterative process (Figure 3.12). If the decomposition reaction is a first order process then the extent of reaction is given by:

$$d\alpha/dt = k\,(1-\alpha) \tag{3.3}$$

Then it is possible to predict the lifetime before *e.g.* 10% mass loss as a function of temperature as shown in Figure 3.13.

Two cautions regarding the use of lifetime prediction:

1. The extent of degradation used in this example (10%) may be unacceptable in terms of its impact on other properties of the material *e.g.* mechanical properties of the solidified polymer.[48]

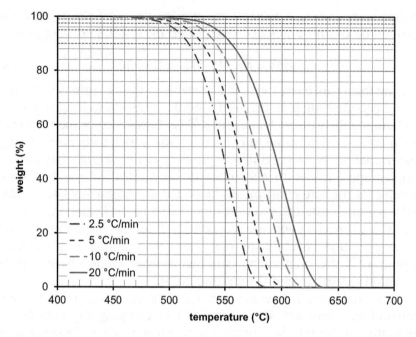

Figure 3.11   Mass loss curves for PTFE at different heating rates.

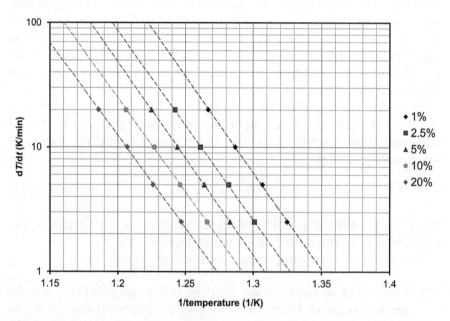

Figure 3.12   Kinetic analysis of results from Figure 3.11 to determine E_a.

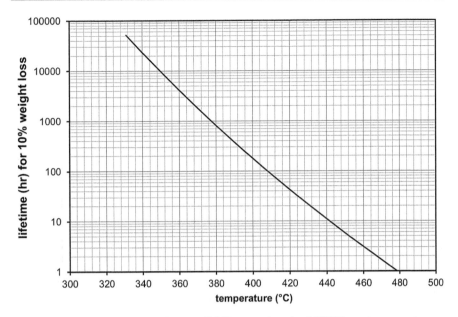

Figure 3.13 Predicted lifetime (10% mass loss) of PTFE *vs.* temperature.

2. These algorithms have been incorporated into a number of commercially available software packages;[47] the user should always question the predictions of such "black box" methods, especially since the assumption that the process follows first order kinetics may not be valid.

The kinetic analysis of thermogravimetric data is a veritable "minefield";[49] the reader is referred to the excellent review of this area by Galwey and Brown[50] for a more in-depth discussion. The recommendations of the ICTAC Kinetics Committee also provide a useful guide for the treatment of thermal analysis data.[51]

3.4 Advanced Temperature Programmes

In the preceding examples, relatively simple heating profiles have been considered whereby the sample temperature is increased in a linear fashion, perhaps with one or more predetermined constant temperature segments to enable a reaction to go to completion (possibly accompanied by a change of sample atmosphere). Whilst the majority of measurements are of this type, there may be benefits obtained by employing more sophisticated temperature programmes.

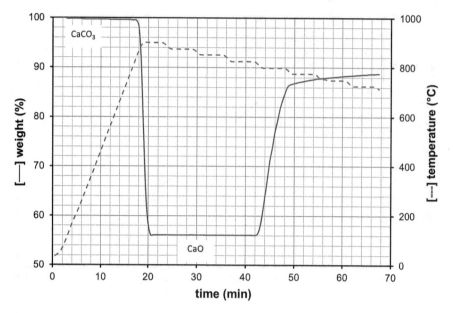

Figure 3.14 Calcium carbonate heated under a humidified atmosphere of 4% CO₂ in nitrogen.

By way of an introduction to this, the use of a stepped reduction in temperature to investigate lime burning is shown in Figure 3.14. In this experiment, calcium carbonate is heated in a humidified mixture of 4% carbon dioxide in nitrogen to simulate the combustion products of burning coke in air. The calcium oxide thus formed is then cooled gradually in a series of isothermal steps 25 °C apart and the re-formation of calcium carbonate observed. The temperature range at which this occurs provides a good indication of the optimum operating temperature for a limekiln.

3.4.1 Sample-controlled Thermogravimetry

In many cases, the mass loss profiles of materials are complex and consist of several overlapping processes. If the heating rate is reduced, it is usually possible to obtain more complete separation of each stage of the decomposition process. Such an approach naturally increases the time for the experiment to be performed. An alternative method is to enable the heating rate to be governed by a property of the sample itself. The generic name for this approach is sample controlled thermal analysis (SCTA), and sample controlled thermogravimetry (SCTG) is now the most widely used SCTA technique. This is discussed in detail in Chapter 11.

3.4.2 Parameter-jump Methods

The use of experiments carried out at a range of heating rates to determine kinetic parameters such as E_a and A, have been described earlier. This method suffers from the disadvantage of requiring several measurements and can be somewhat time consuming especially at the slowest heating rates used. An alternative approach is to perform a single experiment whereby instead of increasing the temperature in a linear fashion, the temperature is increased in a series of steps and the rates of mass loss are extrapolated to the start of the temperature jump between isothermal plateaus.[52–55] This gives $-\mathrm{d}m/\mathrm{d}t$ at the two isothermal plateau temperatures $(T_1$ and $T_2)$ from which E_a may be obtained:

$$E_a = R \left[\ln \left(\frac{\left\{ \frac{\mathrm{d}m}{\mathrm{d}t}(T_1) \right\}}{\left\{ \frac{\mathrm{d}m}{\mathrm{d}t}(T_2) \right\}} \right) \right] \left(\frac{1}{T_1} - \frac{1}{T_2} \right) \qquad (3.4)$$

An experiment of this type concerning the formation of calcium carbonate has already been discussed although the wide spacing and long duration of the temperature steps do not lend themselves to a proper analysis. A better example is illustrated in Figure 3.15 for a sample of benzoic acid undergoing sublimation. Sublimation and

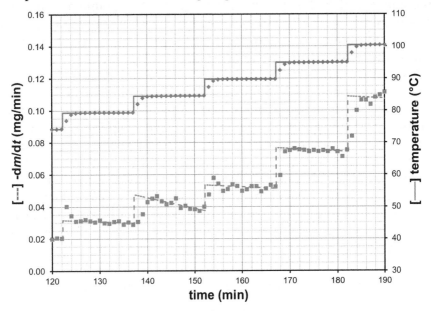

Figure 3.15 Temperature-jump experiment with benzoic acid. The temperature is increased in 5 °C steps every 15 minutes. Lines show extrapolated steps from raw data points.

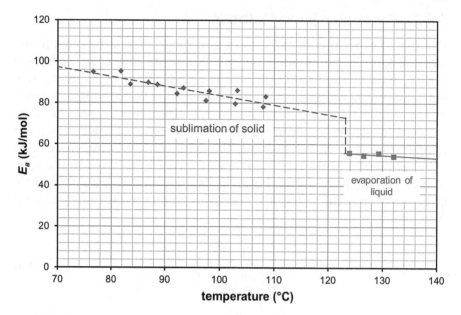

Figure 3.16 E_a values derived from temperature-jump experiments.

evaporation are simple zero-order processes and the rate of mass loss should be constant at any particular temperature providing that the exposed surface area does not change.[56] Although it is sometimes possible to satisfy this condition through careful sample preparation, this requirement can be circumvented by extrapolating the mass loss profiles to obtain an instantaneous change in rates at the point at which the temperature-jump occurred.

From this analysis, the enthalpies of sublimation can be obtained,[57] as illustrated in Figure 3.16 for benzoic acid where the simple analysis described above approximates to the enthalpy of sublimation (or vaporisation if the sample is liquid), while the magnitude of the discontinuity in E_a at the melting temperature of benzoic acid (122 °C) corresponds to the enthalpy of fusion of the material. Thermogravimetry is popular for the study of evaporation and sublimation, although there is considerable debate over the analysis of data.[58]

Rather than change the temperature in steps, it is equally possible to use SCTG to change the rate of mass loss between two set levels. This rate-jump technique is described in Chapter 11.

3.4.3 Modulated-temperature Thermogravimetry

One disadvantage with the temperature-jump method outlined is the requirement to be able rapidly to heat (or cool) the furnace temperature. Furthermore, the data require extensive post-processing in order to determine E_a by extrapolation from the isothermal steps. In order to overcome these drawbacks, a sinusoidal temperature profile can be so that the heating rate is continuously and smoothly alternating.[59] This technique is known as "modulated-temperature thermogravimetry" and allows continuous calculation of E_a during the experiment according to:[60]

$$E_a = \frac{R(T_{av.}^2 - (0.5T_{amp})^2)L}{T_{amp}} \tag{3.5}$$

where $T_{av.}$ is the average thermodynamic temperature, T_{amp} is the amplitude of the temperature modulation and L is the logarithm of the amplitude of the rate of mass loss over one modulation. An example of the raw data from heating a poly(ethylene-*co*-vinyl acetate) copolymer is shown in Figure 3.17. The plot of E_a as a function of conversion in Figure 3.18 distinguishes between side chain and main chain scission as indicated. Although based on the temperature-jump method described above, modulated-temperature thermogravimetry promises to simplify the acquisition of kinetic data.[61–63]

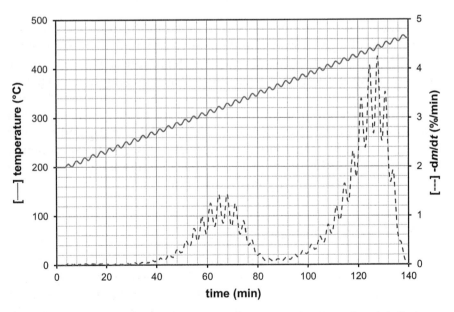

Figure 3.17 Modulated-temperature thermogravimetry of poly(ethylene-*co*-vinyl acetate) under nitrogen, heating rate 1 °C min^{-1}, modulation amplitude 5 °C, period 200 s.

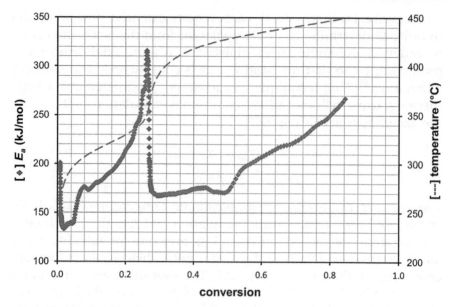

Figure 3.18 Plot of E_a and temperature *vs.* conversion for data from Figure 3.17.

3.5 Simultaneous and Hyphenated Methods

Thermogravimetry cannot be used in isolation for the thermal characterisation of a given system and needs to be supplemented by other thermal methods and general analytical techniques. There are many advantages in carrying out such measurements simultaneously (*i.e.* on the same sample at the same time under the same conditions) and/or concurrently (so that the two measurements can be related to one another) so that complex processes can be investigated without adding uncertainties introduced by performing separate experiments. Further detail of the range of hyphenated methods including those based on thermogravimetry is given in Chapter 10.

3.5.1 Thermogravimetry-DTA/DSC

Differential scanning calorimetry monitors the difference in heat flow between a sample and inert reference substance during a temperature programme. Processes that result in an energy change in the material can be detected and quantified. The technique is described in Chapter 5. Various methods are used to measure the heat flux, but a common arrangement is to place the sample and reference holders on a metal or ceramic plate containing thermocouples that serve to detect the difference in temperature between two containers and convert this to a thermal energy difference. Several instruments have been designed that incorporate the heat flux plate design into a thermobalance, affording a means of performing simultaneous DSC and TG.[64,65] There are several advantages to this approach over conventional DSC and TG notwithstanding the saving in experimental time needed to acquire two sets of data. It is useful to be able to determine energy changes associated with thermal decomposition reactions and the thermogravimetric data can be used to synthesise a suitable baseline against which the calorimetric curve can be corrected.[66]

A simple example of TG-DSC is shown in Figure 3.19 for a flame retardant. The DSC curve clearly shows the melting of the material between 90 and 100 °C and exothermic decomposition of the substance between 230 and 260 °C. The TG curve indicates that some mass loss occurs below 230 °C and closer inspection of the DSC curve suggests that there is an endothermic change also occurring in this region (perhaps even beginning as low as 150 °C). There is no change in mass accompanying the melting of the material and thus this process is "invisible" to thermogravimetry. In some materials, occasionally there is a small mass loss on melting (or at the glass

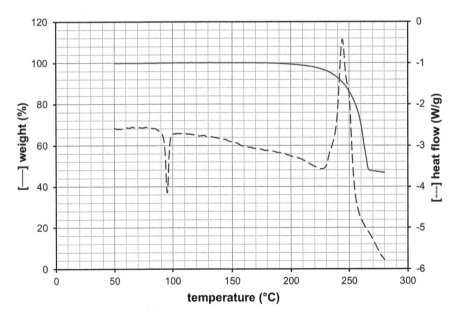

Figure 3.19 Simultaneous TG-DSC experiment on flame retardant, 5 °C min^{-1} under nitrogen.

transition) arising from volatilisation of water or retained solvent whilst other substances degrade on melting. The benefits of combined TG-DSC are obvious in such circumstances.

3.5.2 Thermogravimetry-evolved Gas Analysis

The capability of thermogravimetry for material characterisation is greatly increased if other techniques are coupled to the thermobalance in order to identify the products evolved during the experiment. Non-specific methods such as coupled thermogravimetry-photometry afford a means of measuring smoke generation during the degradation of polymers,[5] however the most usual means of identification of evolved gases employs either infrared or mass spectrometry.[67-69] Both thermogravimetry-infrared (TG-IR) and thermogravimetry-mass spectroscopy (TG-MS) evolved gas analysis use a heated transfer line to carry the exhaust from the thermobalance into the spectrometer in order to prevent the condensation of less volatile products. For TG-IR, a gas flow cell through which the beam from the spectrometer is passed is used. This must also be heated in order to prevent contamination of the optics, which use mirrors to increase the path length of the cell so as to maximise its sensitivity. Mass spectrometry is an alternative to infrared analysis although the interface design is complicated by the requirement to operate the mass spectrometer under high vacuum. Various splitter designs have been developed in order to reduce the transfer line pressure down to a level suitable for injection into the mass spectrometer.[70] Again, these units are heated so as to prevent condensation of less volatile products.

An example of the use of TG-IR is given in Figure 3.20, which shows the mass loss and derivative mass loss profiles for thiophene fulgide.[71] The total signal from the infrared detector is shown in Figure 3.21. Whilst the derivative mass loss curve in Figure 3.20 shows only a broad peak, the infra-red signal exhibits two overlapping peaks, which can be separated by summing the infrared absorbance over a narrower frequency range corresponding to the parent material ($1797-1759 \, \text{cm}^{-1}$) and a strongly absorbing nitrile component ($2306-2206 \, \text{cm}^{-1}$). The latter peak was attributed to decomposition products of the residual dehydrating agent (N,N'-dicyclohexylcarbodiimide) used in its preparation.

An example of the application of TG-MS for catalysis studies is shown in Figure 3.22 for a sample of acid-treated clay that was infused with propanol vapour.[72] The TG curve shows the loss of volatiles from the clay and the selective ion currents corresponding to the evolution

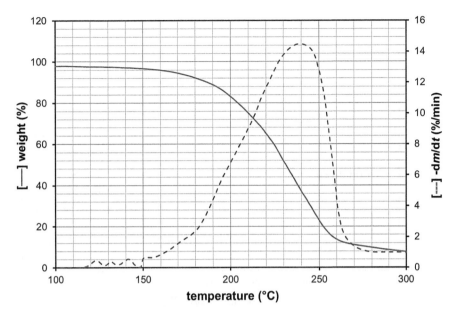

Figure 3.20 Mass and rate of mass loss *vs.* temperature for thiophene fulgide, 10 °C min^{-1} in nitrogen.

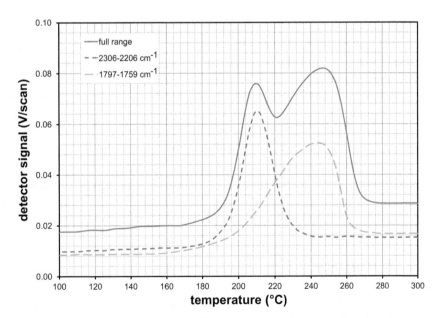

Figure 3.21 Total and selected IR absorption of evolved gases from experiment shown in Figure 3.20.

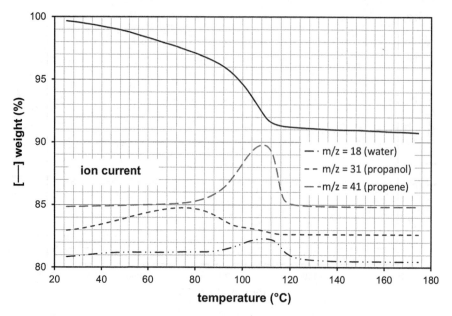

Figure 3.22   TG-MS experiment on clay infused with propanol.

of propanol (as $^{+\bullet}CH_2OH$ fragment, $m/z = 31$), water ($m/z = 18$) and propene ($^{+\bullet}CH_2CH=CH_2$, $m/z = 41$). As the temperature is increased, the adsorbed alcohol is dehydrated to the corresponding alkene. The yield and temperature of propene evolution can be used to determine the most effective methods of activating the substrate.[73]

Although both infrared and mass spectrometry are valuable techniques for the identification of evolved gases, in many cases a mixture of decomposition products is evolved simultaneously. Several workers have added the capability to perform offline[74,75] or online[76,77] gas chromatography to separate the effluent into individual components, which can then be identified by mass spectrometry or infrared spectroscopy. Such approaches[78,79] essentially attach an analytical laboratory to the end of the humble thermobalance and make this an appropriate point to conclude this introduction to the technique.

3.6 Summary

Thermogravimetry is a technique commonly used for the determination of solvent content and the analysis of degradation and reaction processes. The most widespread applications are compositional analysis and thermal stability, although with proper consideration of experimental design, a more sophisticated interpretation can be performed. As TG alone does not identify chemical species, it is important to take care in assigning mass losses in the absence of further analysis. Hyphenated and sample-controlled TG add further scope to this method and are described in Chapters 10 and 11, respectively.

References

1. C. J. Keattch, *An Introduction to Thermogravimetry*, Heyden & Son, London, 1969.
2. G. Urbain and Ch. Boulanger, *Compt. Rend.*, 1912, **154**, 347.
3. K. Honda, *Sci. Rep. Tohoku Univ.*, 1915, **4**, 97.
4. C. Duval, *Inorganic Thermogravimetric Analysis*, Elsevier, 2nd edn, 1963.
5. J. Chiu, *Appl. Polym. Symp.*, 1966, **2**, 25.
6. L. Cahn and H. E. Schultz, *Vacuum Microbalance Techniques*, ed. R. F. Walker, Plenum Press, New York, 1962, **2**, pp. 7–18.
7. A. Kettrup, G. Matuschek, H. Utschick, Ch. Namendorf and G. Bräuer, *Thermochim. Acta*, 1997, **295**, 119.
8. A. Skreiberg, Ø. Skreiberg, J. Sandquist and L. Sørum, *Fuel*, 2011, **90**, 2182.
9. F. Dreisbach and H. W. Lösch, *J. Therm. Anal. Calorim.*, 2000, **62**, 515.
10. R. E. Latta, J. T. Bittel and G. B. Hadesty, *Rev. Sci. Instrum.*, 1967, **38**, 1667.
11. J. Cuya, N. Sato, K. Yamamoto, H. Takahashi and A. Muramatsu, *High Temp. Mater. Processes*, 2003, **22**, 197.
12. ASTM E2040 – 08 (2014) *Standard Test Method for Mass Scale Calibration of Thermogravimetric Analyzers*, ASTM International, West Conshohocken, 2013.
13. R. P. Tye, R. L. Gardner and A. Maesono, *J. Therm. Anal.*, 1993, **40**, 1009.
14. E. Pedersen, *J. Phys. E: Sci. Instrum.*, 1968, **1**, 1013.
15. S. Hoffmann, M. Schmidt, S. Scharsach and R. Kniep, *Thermochim. Acta*, 2012, **527**, 204.
16. J. R. Williams and W. W. Wendlandt, *Thermochim. Acta*, 1973, **7**, 253.
17. S. Dobner, G. Kan, R. A. Graff and A. M. Squires, *Thermochim. Acta*, 1976, **16**, 251.

18. H. G. Wiedemann, *Vacuum Microbalance Techniques*, ed. C. H. Massen and H. J. van Beckum, Plenum Press, New York, 1970, p. 217.
19. S. Indrijarso, J. S. Oklany, A. Millington, D. Price and R. Hughes, *Thermochim. Acta*, 1996, **277**, 41.
20. G. Lopez, R. Aguado, M. Olazar, M. Arabiourrutia and J. Bilbao, *Waste Manage.*, 2009, **29**, 2649.
21. A. R. McGhie, J. Chiu, P. G. Fair and R. L. Blaine, *Thermochim. Acta*, 1983, **67**, 241.
22. ASTM E1582 - 14 *Standard Practice for Calibration of Temperature Scale for Thermogravimetry*, ASTM International, West Conshohocken, 2014.
23. S. D. Norem, M. J. O'Neill and A. P. Gray, *Thermochim. Acta*, 1970, **1**, 29.
24. L. N. Stewart, *Proc. 3rd Toronto Symp. Thermal Anal.*, ed. H. G. McAdie, Chemical Institute of Canada, Toronto, 1969, 205.
25. P. Le Parlouër, *Thermochim. Acta*, 1985, **92**, 371.
26. A. E. Newkirk, *Thermochim. Acta*, 1971, **2**, 1.
27. J. P. Sanders and P. K. Gallagher, *Thermochim. Acta*, 2003, **406**, 241.
28. S. St J. Warne and P. K. Gallagher, *Thermochim. Acta*, 1987, **110**, 269.
29. P. K. Gallagher, *J. Thermal Anal. Calorim.*, 1997, **49**, 33.
30. P. Kamasa and P. Myśliński, *Thermochim. Acta*, 1999, **337**, 51.
31. D. M. Lin, H. S. Wang, M. L. Lin, M. H. Lin and Y. C. Wu, *J. Therm. Anal. Calorim.*, 1999, **58**, 347.
32. D. Thickett and M. Odlyha, *J. Therm. Anal. Calorim.*, 2005, **80**, 565.
33. A. Schiraldi and D. Fessas, *J. Therm. Anal. Calorim.*, 2003, **71**, 225.
34. C. M. Ernest, *Compositional Analysis by Thermogravimetry*, ASTM STP 997, American Society for Testing and Materials, Philadelphia, 1998.
35. E. L. Charsley, J. A. Rumsey, S. B. Warrington, J. Robertson and P. N. A. Seth, *Thermochim. Acta*, 1984, **72**, 251.
36. ASTM D7582 - 12 *Standard Test Methods for Proximate Analysis of Coal and Coke by Macro Thermogravimetric Analysis*, ASTM International, West Conshohocken, 2012.
37. C. J. Donahue and E. A. Rais, *J. Chem. Educ.*, 2009, **86**, 222.
38. K. J. Laidler, *J. Chem. Educ.*, 1984, **61**, 491.
39. J. Šesták and G. Berggren, *Thermochim. Acta*, 1971, **3**, 1.
40. M. E. Brown, *Introduction to Thermal Analysis*, Kluwer, Dodrecht, 2nd edn, 2001.
41. M. E. Brown and A. K. Galwey, *Thermochim. Acta*, 1976, **29**, 129.

42. J. H. Flynn, *J. Thermal Anal. Calorim.*, 1995, **44**, 499.
43. T. Ozawa, *Bull. Chem. Soc. Jpn.*, 1965, **38**, 1881.
44. J. H. Flynn and L. A. Wall, *Polym. Lett.*, 1966, **4**, 323.
45. ASTM E1641-13 *Standard Test Method for Decomposition Kinetics by Thermogravimetry*, ASTM International, West Conshohocken, 2013.
46. ISO 11358-3:2013 *Thermogravimetry (TG) of polymers - Part 3: Determination of the activation energy using the Ozawa-Friedman plot and analysis of the reaction kinetics*, International Organization for Standardization, Geneva, 2014.
47. J. Opfermann and E. Kaisersberger, *Thermochim. Acta*, 1992, **203**, 167.
48. ASTM E1877 - 13 S*tandard Practice for Calculating Thermal Endurance of Materials from Thermogravimetric Decomposition Data*, ASTM International, West Conshohocken, 2013.
49. M. E. Brown, *J. Therm. Anal. Calorim.*, 1997, **49**, 17.
50. A. K. Galwey and M. E. Brown, Handbook of Thermal Analysis and Calorimetry, in *Principles and Practice*, ed. M. E. Brown, Elsevier Science B. V., Amsterdam, 1998, vol. 1., pp. 147–224.
51. S. Vyazovkin, A. K. Burnham, J. M. Criado, L. A. Pérez-Maqueda, C. Popescu and N. Sbirrazzuoli, *Thermochim. Acta*, 2011, **520**, 1.
52. T. Akahira, *Sci. Pap. Inst. Phys. Chem. Res.*, 1928, **9**, 165.
53. J. H. Flynn and B. Dickens, *Thermochim. Acta*, 1976, **15**, 1.
54. B. Dickens, *J. Polym. Sci., Polym. Chem. Ed.*, 1982, **20**, 1065.
55. B. Dickens, *J. Polym. Sci. Polym. Chem. Ed.*, 1982, **20**, 1169.
56. S. J. Ashcroft, *Thermochim. Acta*, 1972, **2**, 512.
57. D. M. Price, S. Bashir and P. R. Derrick, *Thermochim. Acta*, 1999, **327**, 167.
58. S. P. Verevkin, R. V. Ralys, D. H. Zaitsua, V. N. Emel'yanenko and C. Schick, *Thermochim. Acta*, 2012, **538**, 55.
59. J. H. Flynn, *Thermal Analysis, Proc. ICTA 2*, ed. R. F. Schwenker Jr. and P. D. Garn, Academic Press, Worcester, 1969, pp. 1111–1126.
60. R. L. Blaine and B. K. Hahn, *J. Thermal Anal*, 1998, **54**, 695.
61. V. Mamleev and S. Bourbigot, *Chem. Eng. Sci.*, 2005, **60**, 747.
62. V. Mamleev, S. Bourbigot, M. Le Bras, J. Yvon and J. Lefebvre, *Chem. Eng. Sci.*, 2006, **61**, 1276.
63. N. Koga, Y. Goshi, M. Yoshikawa and T. Tatsuoka, *J. Chem. Educ.*, 2014, **91**, 239.
64. Q. Lineberry and W.-P. Pan, *Simultaneous Techniques Including Analysis of Gaseous Products, Characterization of Materials*, 2012, pp. 1–14.

65. E. L. Charsley, in *Thermal Analysis - Techniques & Applications*, ed. E. L. Charsley and S. B. Warrington, The Royal Society of Chemistry, Cambridge, 1992, pp. 59–83.
66. K. Sigrist and H. Stach, *Thermochim. Acta*, 1996, **278**, 145.
67. S. B. Warrington, in *Thermal Analysis - Techniques & Applications*, ed. E. L. Charsley and S. B. Warrington, The Royal Society of Chemistry, Cambridge, 1992, pp. 84–107.
68. S. Materazzia and S. Vecchio, *Appl. Spectrosc. Rev.*, 2013, **48**, 654.
69. S. Materazzia and S. Vecchio, *Appl. Spectrosc. Rev.*, 2011, **46**, 261.
70. G. Szekely, M. Nebuloni and L. F. Zerilli, *Thermochim. Acta*, 1992, **196**, 511.
71. D. M. Price, S. P. Church and C. P. Sambrook-Smith, *Thermochim. Acta*, 1999, **332**, 197.
72. E. L. Charsley, C. Walker and S. B. Warrington, *J. Thermal Anal. Calorim.*, 1993, **40**, 983.
73. A. De Stefanis and A. A. G. Tomlinson, *Catal. Today*, 2006, **114**, 126.
74. W. H. McClennen, R. M. Buchanan, N. S. Arnold, J. P. Dworzanski and H. L. C. Meuzelaar, *Anal. Chem.*, 1993, **65**, 2819.
75. T. J. Lever, D. M. Price and S. B. Warrington, *Proc. Conf. North Am. Therm. Anal. Soc. 28th*, 2000, 4.
76. H. L. Chung and J. C. Aldridge, *Instrum. Sci. Technol.*, 1992, **20**, 123.
77. E. C. Sikabwe, D. L. Negelein, R. Lin and R. L. White, *Anal. Chem.*, 1997, **69**, 2606.
78. M. Webb, P. M. Last and C. Breen, *Thermochim. Acta*, 1999, **326**, 151.
79. C. Breen, P. M. Last, S. Taylor and P. Komadel, *Thermochim. Acta*, 2000, **363**, 93.

4 Dynamic Vapour Sorption

Nicole Hunter

Thermo Fisher Scientific Inc., Basingstoke, UK
Email: Nicole.Hunter@Thermofisher.com

4.1 Introduction and Principles

Dynamic vapour sorption (DVS) is a gravimetric technique used to measure the mass loss/gain of a material as a function of humidity. It is possible to use organic solvents on many commercially available instruments but for ease of explanation, water vapour will be referred to as the norm throughout the chapter. The most common experiment carried out using DVS analysers is the construction of sorption isotherms for materials but the technique can be used for a wide range of applications across several industries including food, pharmaceuticals, textiles, building materials and explosives. The raw data are usually presented as normalised mass percentage *versus* time. Sorption isotherms are then conventionally presented as equilibrium water content *versus* relative humidity.

The ability of a sample to take on or release moisture in a specific atmosphere is a critical attribute when characterising a material. It dictates the conditions required for manufacturing, transportation and storage and must be understood if a product from any of the aforementioned industries is to be made successfully and reproducibly. Because DVS analysis has a wide range of applications, the technique has grown in popularity over the past ten years and it now features in the majority of laboratories focussing on material characterisation.

Principles of Thermal Analysis and Calorimetry: 2nd Edition
Edited by Simon Gaisford, Vicky Kett and Peter Haines
© The Royal Society of Chemistry 2016
Published by the Royal Society of Chemistry, www.rsc.org

The relationship between humidity and temperature is well known and it is crucial to control one of these in order to study the effect of the other. Relative humidity (% RH) is intrinsically dependent on the temperature of the air/gas containing the water vapour. For example, a kilogram of air is capable of holding a maximum mass of water vapour at a specific temperature (without condensation occurring), but if the temperature of the air is increased and the quantity of water is kept constant, the relative humidity will decrease due to the expansion of the air at the higher temperature. This means that any experiments that are designed to investigate the effect of humidity on a material must have very precise temperature control to ensure that measurements are truly being made under the conditions reported. The majority of DVS experiments are carried out isothermally in order to avoid changing two variables at the same time (although it is also possible to carry out experiments at a constant humidity and vary the temperature).

Control of temperature is extremely important because it is well known that, for most reactions, the reaction rate almost always varies with temperature.[1] A very small temperature fluctuation can cause a marked change in the rate of a reaction. It is therefore of paramount importance that the measurements are made at a very tightly controlled temperature (or temperature program) to ensure consistency and accuracy of results.

Given that the International Confederation for Thermal Analysis and Calorimetry (ICTAC) definition of thermal analysis is the 'study of the relationship between a sample property and its temperature as the sample is heated or cooled in a controlled manner',[2] it is appropriate that DVS is included in the group of thermal analysis techniques. DVS monitors the mass of a sample as the temperature and/or humidity are held and/or varied throughout the experiment.

A fairly recent addition to DVS instruments is the option to acquire additional data during experiments through use of a Raman[3] or near infrared (NIR) probe[4] or a microscope. Like similar hyphenated techniques such as DSC-Raman, this offers chemical or visual data to aid in the interpretation of mass changes for a sample rather than relying on the mass data alone.

4.2 Definitions and Nomenclature

Vapour pressure (or equilibrium vapour pressure) can be described as the pressure exerted by a vapour when in equilibrium with the liquid phase in a closed system at a defined temperature. The vapour pressure of a substance is dependent on the temperature of the system and as the temperature is raised, the vapour pressure will increase.

Relative humidity (% RH) is commonly used to express the water content of the air in an experiment. This is simply the ratio of the actual vapour pressure of the water vapour compared to the saturation vapour pressure, eqn (4.1).

$$\text{Relative humidity:} \frac{\text{partial vapour pressure } (P)}{\text{saturation vapour pressure } (P_0)} \times 100\% \qquad (4.1)$$

This gives a representation of the amount of water in the air compared with the maximum possible at that temperature. A relative humidity of 100% therefore represents complete saturation of the air with water vapour and at all other relative humidity values, the air is below the saturation point.

Vapour sorption analysts may also refer to water activity (a_w) when reporting results. Water activity is defined as the effective mole fraction of water in air and is equal to the relative humidity expressed as a fraction (*e.g.* 90% RH is equivalent to a water activity of 0.9). It is represented by the relationship shown in eqn (4.2), where P and P_0 are the actual vapour pressure of water in air and the saturation vapour pressure, as before:

$$\text{Water activity, } a_w = \frac{P}{P_0} \qquad (4.2)$$

The water content of the sample may also be represented in different ways. The data produced by the instrument is usually either a mass percentage of the initial sample mass or a change in mass as a percentage; this can then be used to obtain a percentage mass of water in the sample during the experiment (*e.g.* 3% water (mass/mass)). It may also be an actual mass or mass change in mass units (usually milligrams but may be grams for experiments using larger samples). It is also common to see moisture content expressed as a molar ratio (*e.g.* 1.2 moles of water per mole of the substance) in order to provide some physical meaning for the data.

4.3   Principles of the Technique

The main principle of DVS is that as a sample is exposed to a specific relative humidity, the sample will either take on or lose water with

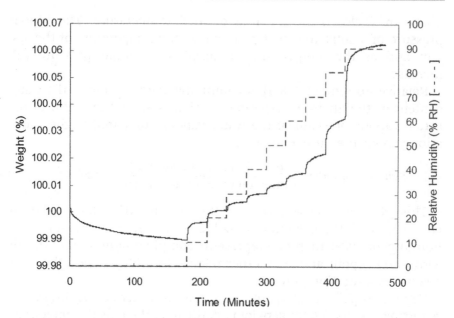

Figure 4.1 Sorption isotherm data for crystalline trehalose dihydrate.

time and this is observed as a mass change. Experiments are carried out with accurate temperature control as this directly affects the sorption behaviour of the sample. A common DVS experiment is the generation of a sorption isotherm for a material; the humidity is incrementally increased (and/or decreased) at a constant temperature and the equilibrium water content of the material is recorded. An example of typical sorption isotherm data from a DVS analyser is shown in Figure 4.1.

Before the advent of the high accuracy instruments currently available, humidity controlled experiments were generally conducted using saturated solutions of specific salts in order to generate the desired humidity within an enclosed container. The relative humidity inside a closed vessel containing pure water will reach 100% saturation when the liquid phase is fully equilibrated with the vapour phase. That is, water molecules will continue to move into the vapour phase until the atmosphere is at the maximum possible water content without condensation occurring. Water molecules continuously move between the liquid and vapour phase in order to maintain the equilibrium. In the case of saturated salt solutions, the proportion of water molecules in the solution is clearly lower than for pure water. This reduces the number of water molecules available to move into the vapour phase and slows the rate of evaporation. The rate of the return of water vapour back into the liquid phase remains unaltered though giving rise to a reduced amount of water in the vapour phase overall. All saturated salt solutions therefore generate a lower relative humidity than pure water *i.e.* less than 100% RH. As an example, a saturated solution of sodium chloride at 25 °C will generate a relative humidity of 75.3% whereas a saturated solution of lithium chloride at the same temperature will provide a relative humidity of 11.3%.[5]

There are, however, several practical problems and pitfalls with this kind of experimental set up. Firstly, the analyst must be absolutely certain that a saturated solution has been created or the humidity will be higher than that intended (for the reasons explained earlier). Secondly, the temperature must be very carefully maintained using an independent source of heat such as an oven or a fridge; most standard lab equipment is not capable of maintaining the level of precision required to make accurate measurements. For example, an increase in temperature from 25 to 30 °C changes the relative humidity of a saturated solution of potassium fluoride from 30.8 to 27.3% RH.[5] Finally, the major problem with this type of experiment is that the controlled atmosphere usually has to be disturbed to measure the sample response *i.e.* the sample has to be removed from the system to weigh it. This is problematic for many reasons:

1. The air inside the chamber will be disturbed and will have to re-equilibrate once the chamber is closed again. Depending on the size of the chamber, this can take a significant amount of time in relation to the length of the experiment and means that the sample is not exposed to the desired humidity consistently throughout the experiment.
2. Sample handling may be challenging as it may lose/gain moisture between removing it from the chamber and weighing it, making the result inaccurate. This can happen extremely quickly for hygroscopic materials. Handling under an inert atmosphere (such as dry nitrogen) may help to overcome this but is often not a solution as the sample is likely to lose moisture to the drier atmosphere.
3. The experiments can be labour intensive depending on the frequency of sampling required. This is especially problematic where several different combinations of temperature and humidity are required for the study.
4. It is extremely difficult to make real-time measurements using this type of approach. When samples are taken for measurement, there will be a time delay between removing the sample

and acquiring the result. Again, this is likely to lead to inaccurate results, especially where kinetic information (rate of absorption, diffusion *etc.*) is required.

5. Experiments can be very long due to the sample mass and chamber size, making equilibration times much longer than on a DVS instrument. An additional issue can be mould growth for high humidity environments with some samples due to the timings involved.

6. The chamber contains a static atmosphere of the required humidity; a key feature of DVS analysers is that the humidified air is constantly flowed over the sample rather than being static. This carries away any contaminant vapours from the sample and reduces the effects of localised micro-environments close to the exposed surface of the sample.

That is not to say, however, that this type of experiment has no place today. The main point of note here is that where accurate kinetic data are required, the only practical option is to use a DVS instrument.

By far the most common type of DVS experiment is the generation of a moisture (or other vapour) sorption isotherm for a material at a particular temperature (as shown in Figure 4.1). This characterises the behaviour of a sample over the entire humidity range and as such can be used to determine stability during storage and transportation of a product and can be used to predict shelf-life with respect to humidity. Sorption isotherms are also useful for determining the kinetics of vapour sorption and vapour diffusion coefficients under specified environmental conditions. Multiple sorption isotherm experiments may be carried out to study the behaviour over a range of temperatures as well as humidity.

Sorption isotherm experiments expose the sample to a series of humidity steps at a constant temperature and the sample is allowed to come to equilibrium with the applied humidity during each step. The moisture content of the material when in equilibrium with each applied humidity is plotted as a function of RH. This generates a sorption isotherm (Figure 4.2).

Sorption isotherms can be generated simply by increasing the humidity but in order to characterise a material properly, the behaviour should also be studied when the humidity is decreased in the same size steps. Most materials do not release water at the same rate as they adsorb water and the difference in water vapour uptake between the sorption and desorption isotherms is termed the hysteresis of the material. Very similar sorption and desorption curves (*i.e.* little to no

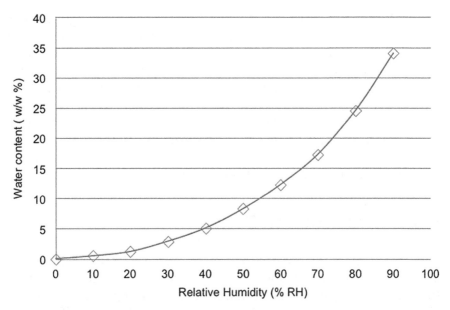

Figure 4.2 A schematic example of a sorption isotherm.

hysteresis) indicate that the material adsorbed water onto the surface in the increasing humidity run and then the water simply desorbed from the surface at a similar rate in the decreasing humidity part of the experiment. Where the sorption and desorption curves are different, this indicates that the material absorbed water into the structure during the increasing humidity steps. As the rate of release of the water out of the structure is usually slower than the absorption into it, this gives rise to a difference in the absorption/desorption profiles. If the material is wetted again after the desorption steps and the second absorption curve does not match the first one, this indicates that the material has been physically changed during the experiment. For example, it may have changed its hydration state, crystalline structure or have hydrolysed.

Materials can be classified according to the shape of their adsorption isotherms. Most materials fall into one of six classic categories. A full description of these and their significance is given by IUPAC.[6] These models can be useful in understanding the physical processes occurring for a material during the humidity program and can also reveal information about the pore size for a material.

Several adsorption isotherm models exist to characterise the behaviour of a material but the most commonly used are the Brunauer–Emmett–Teller (BET) and Guggenheim–Anderson–de Boer (GAB).

The GAB model is often the most useful for characterising materials since the BET model is only appropriate for water activities up to 0.50 a_w. The models are useful in predicting the adsorbed moisture content of a material at a given water activity but are not useful for materials where absorption occurs due to the fundamental principles of these models (*i.e.* that the water molecules create layers on the surface of the material and do not move into the structure). Explanations of these models are freely available in the literature and are beyond the scope of this book so are not included here but Foo and Hameed[7] provide an excellent review.

4.4   Instrumentation Design

The exact capabilities of any particular DVS analyser will by stated by the manufacturer. Values given here for instrument capabilities (humidity range, temperature range *etc.*) are an approximation and should be treated as a guide. In order to ensure that a specific model is capable of performing the exact experiment required, the manufacturer should be consulted directly.

There are two major components of a dynamic vapour sorption instrument: a temperature and humidity controlled chamber and a highly sensitive microbalance. Sample and reference crucibles are contained within the humidity chamber and are connected to the balance by means of hang-down wires. As the sample and reference materials are exposed to the required environmental conditions, their masses are measured in real time. The crucibles are grounded to eliminate static, which might otherwise cause a significant problem for DVS measurements. Selection of an appropriate crucible material can also help to reduce static issues (quartz, for example, rather than platinum). The use of a reference crucible means that any conditions affecting the crucible itself (surface wetting, for example) are removed from the resulting signal, which results in a higher sensitivity. The reference must usually, therefore, be empty and the same crucible type as the one containing the sample. Considerations for selecting an appropriate crucible are discussed in the 'Experimental Considerations' section. Many of the commercially available systems now come with a touchscreen, which provides a user friendly interface for operation, and saves time if the controller PC is not easily accessible. Figure 4.3 shows a schematic of a typical instrument set up.

Nitrogen (or any other appropriate dry gas) is supplied to the humidity chamber *via* two mass flow controllers. One gas line passes

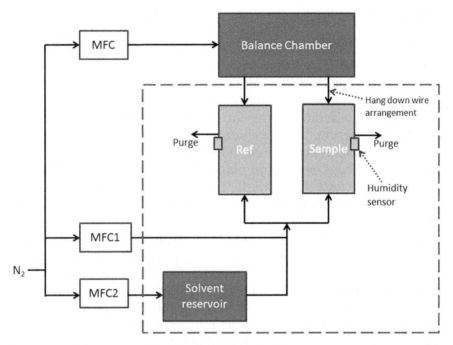

Figure 4.3 A schematic of a typical dynamic vapour sorption analyser. MFC 1, 2 and 3 refer to mass flow controllers (image adapted from IA Instruments Applications Note *TA329a Moisture Sorption Analysis of Pharmaceuticals*. Used with permission).

through a solvent reservoir to achieve 100% saturation of the carrier gas with the vapour of interest (*e.g.* water vapour). The other gas line remains completely dry. The mass flow controllers proportion the amount of saturated and dry gas to achieve the set humidity; the gases are then mixed before entering the sample and reference chambers. The chambers are relatively small, which enables the environment to equilibrate quickly to the desired conditions.

The humidity can usually be set in the region of 0 to 98% RH. For applications requiring extremely dry or humid atmospheres, the system capability should be checked with the manufacturer. Humidity programs may be any of the following:

- Stepwise increments of increasing/decreasing humidity (as in the generation of moisture sorption isotherms)
- Ramped humidity (*i.e.* a linear rate of change in the applied humidity such as 0.2% RH per minute)
- Constant humidity *i.e.* the entire experiment is conducted at one humidity setting

Stepped humidity experiments are by far the most common and steps as small as 0.1% RH can be used. Caution should be exercised here though as the accuracy of some systems is around 1% despite the apparent resolution being much lower.

A common approach for sorption isotherm experiments is to program the instrument to move on to the next humidity when the sample has equilibrated at the current humidity. The analyst sets the criteria of a maximum mass change over a specified time period for when the sample is considered to be fully equilibrated (*e.g.* less than 0.2% mass change over 30 minutes). The instrument will then not move on to the next step until the criteria is met. It is also possible to set a maximum time for each step to avoid unnecessarily long experiments.

There are two common types of analyser: those with an open loop humidity control system and those with a closed loop control system. In both systems, the desired humidity is generated by mass flow controllers and is then directed to the sample chamber. In an open loop system, sensors within the chamber monitor the humidity but do not feed information back to the mass flow controllers. This means that the controllers do not adjust the humidity in relation to the sample behaviour; *i.e.* if a sample releases/absorbs moisture and therefore raises/lowers the humidity in the local environment, the controllers will not adjust the flow to account for this. They will continue to flow the same proportion of saturated and dry air according to the set program. Closed loop systems contain a feedback sensor (commonly a chilled mirror dew point sensor), which actively relays humidity information to the controllers so the humidity can be adjusted according to the behaviour of the sample. The mass flow controllers, therefore, work to maintain the desired humidity depending on the water activity of the sample by adjusting the proportioning of dry and saturated gas. Closed loop systems are therefore suited to direct water activity measurements whereas the open loop systems rely solely on mass changes.

Accurate temperature control of the humidity chamber is vital and the temperature program must be set in addition to the humidity program. Often, experiments are isothermal but it is also possible to create a stepped or ramped temperature program at a constant humidity in the same way that the humidity can be varied. Instruments are typically capable of carrying out experiments at temperatures in the region of 5 to 85 °C although the exact range will depend on the manufacturer; high temperature instruments are now available, which are capable of performing accurate measurements up to 150 °C.

Pre-drying of samples is also possible; the maximum temperature for this stage depends on the instrument but 150 °C is common.

The balances contained within this type of system are highly sensitive and can detect very small changes in the mass of the sample. A sensitivity of 0.1 μg is not uncommon although the capabilities of the systems are different for each of the manufacturers. Dry gas is supplied to the balance to maintain a dry atmosphere inside the balance housing. The balances are usually individually thermostatted so their temperature is independent of the temperature inside the humidity chamber. As balances are very sensitive to heat, this removes some of the issues usually referred to as 'buoyancy' effects and ensures that accurate masses are recorded. The dry gas purge at a constant flow rate also serves to further reduce these effects.

Typical sample masses range from 1 to 500 mg, which vastly reduces the sample equilibration time compared to the experiments carried out in jars of saturated salt solutions. The actual sample equilibration time will depend on the nature of the sample and the humidity being studied but equilibration can often be achieved in a matter of minutes or hours. The exception to this is where samples deliquesce in which case equilibration will not be reached until a saturated solution of the sample is arrived at. Instruments are also available that are specifically designed for larger sample sizes (up to 5 g).

Autosamplers are fitted as standard for some systems. These can be very useful where a large number of samples needs to be run (*e.g.* in a quality control laboratory) or where an experiment finishes at an inconvenient time. Previously, autosamplers have realistically been of limited value for DVS experiments due to the concern that samples could gain or lose moisture whilst in the queue. Modern autosamplers are vastly improved and some now come with the facility to use sealed pans, which can be pierced by the autosampler just before the experiment starts.

Solvents other than water can be used on some (but not all) instruments. Ethanol, for example, is a commonly used alternative to water as it may be carried over from processing steps for some materials so it may be useful to study its effects. Sorption experiments using organic solvents have also been shown to be a useful technique for quantifying low levels of amorphous content in some pharmaceutical systems.[8,9] A range of volatile solvents may be used including other alcohols, hydrocarbons and aromatics but it advisable to check with the manufacturer for any that could be incompatible with the internal components of the system. It is also advisable to ensure that the system has adequate extraction of waste gases where harmful solvents (and samples) are used; it may be necessary to house the instrument inside a fume cupboard or other local exhaust ventilation (LEV) system.

Use of Raman and NIR probes and optical microscopes is becoming increasingly common for DVS experiments. As with the majority of thermal methods of analysis, the signal can be sometimes be difficult to interpret due to the absence of any chemical or visual data. Adding an additional method of analysis can aid in the understanding of unusual/complex events. Manufacturers that offer this option insert a probe inside the humidity chamber so it is focussed on the surface of the sample. The chamber remains a sealed environment so the behaviour of the sample should be unaffected by the presence of the probe.

4.5 Experimental Considerations/Best Practice

When designing the program for an experiment, a number of factors must be considered to ensure that the data generated are going to be useful. The type of program used will be dependent on the nature of the sample and the information required. Will the humidity program be stepped (as for sorption isotherm experiments), or ramped, or will the humidity be held constant throughout the experiment? Whilst stepped programs are very popular, one issue for them is that the material may physically change during the steps preceding the humidity of interest. This could be a crystallisation, which is usually obvious from the data as a concurrent expulsion of vapour normally results, but could also be a more subtle change like a glass transition, which may not be immediately apparent from the data. For applications where information is required on a sample's behaviour in a specific environment, a constant humidity experiment is clearly the recommended approach.

Related to this type of consideration is whether or not to use a sample pre-drying step before the experiment. This is essential for accurate sorption isotherm experiments in order to quote accurate water activity values for the material at very low humidity. Alternatively, the analyst may choose not to pre-dry the sample before a constant humidity experiment in order to study the behaviour of the sample 'as-is'; for example, how does a pharmaceutical tablet behave when exposed to 30% RH? Most analysers are capable of heating the material to an elevated temperature to dry it before changing to the desired experimental temperature. This can remove all water from the sample (free water and chemically bound water) if the temperature and time are carefully selected. Care must be taken not to degrade the sample during this stage though or the resulting DVS data will not represent the behaviour of the pure material. Ideally, thermogravimetric analysis (see Chapter 3) should be performed prior to the DVS experiment to select suitable drying conditions where available. Alternatively, the experiment can be started at 0% RH at the experiment temperature, which will remove any free/ unbound water in the sample if it is allowed to properly equilibrate. This approach is useful for studying hydrate/solvate materials but care should again be taken to ensure that the correct chemical entity is being studied in the subsequent DVS experiment; weakly bonded hydrates can sometimes also be lost at 0% RH but this is tempera- ture dependent.

Sorption isotherm experiments are frequently used for character- ising materials but what size should the humidity steps be? Steps of 5 or 10% relative humidity are often used but the step size selected should depend on the level of detail required for generating the isotherm for that particular material. In practical terms, it may also depend on how much time is available to carry out the experiment; clearly the higher the number of steps in the program, the longer the experiment will take. The humidity range also needs to be defined. For some applications, selection of a narrow range of humidity may be appropriate (*e.g.* study of a material's behaviour in a particular country's climate) but for other applications the entire range should be included (*e.g.* generation of a sorption isotherm). It is also im- portant to ensure that each step is held for an appropriate amount of time. It is possible to set each step to the same time period but it is unlikely that a sample will take the same amount of time to reach equilibration at each humidity studied and the sample must have achieved equilibrium in order to record the accurate equilibrium moisture content for a sorption isotherm. It is therefore usually more useful to set criteria within the program, which prevents the method from advancing to the next step until no further mass change is ob- served. The analyst must decide what constitutes 'equilibrium' and will need to enter stabilisation criteria for the maximum mass change acceptable over a specified time period (*e.g.* the sample mass must be stable to within 0.05% for 30 minutes). Avoid setting criteria though that is beyond the capability of the instrument as all systems have a degree of baseline drift. It is also possible to add a maximum time limit for the steps so that in the event that the sample does not reach equilibration within a reasonable time frame (deliquescence could cause this), the instrument will still move on as instructed after the maximum time has elapsed. The use of less rigorous settings for the equilibration criteria (or a shorter maximum step time) will reduce the experiment time, but may give non-equilibrium results.

The temperature, or temperature program, that will be used is also important to set and must be stated when reporting DVS results. The behaviour of a sample may be quite different at two different temperatures so it is important to select a temperature that is relevant to the application. For example, if the behaviour of a sample during cold shipping is being investigated, it is unlikely to be useful to set the experiment temperature to 25 °C. If the behaviour over a range of temperatures is of interest, there are two options. The first is to carry out the entire experiment several times at a different temperature each time within the desired range. Another option is to fix the humidity and vary the temperature during the experiment so the effect of increasing or decreasing temperature at the same humidity can be studied. It is not usually advisable to change both the humidity and the temperature at the same time in any experiment as it can be very difficult or impossible to ascertain which factor is causing the behaviour in the sample. It is possible though to use a combination of temperature and humidity steps as long as the sample is allowed to equilibrate before the next step commences (*e.g.* initially hold at 20% RH at 25 °C, allow to equilibrate, then step to 30% RH at the same temperature, allow to equilibrate and then step the temperature to 30 °C). This type of experiment can be useful to replicate the transportation of a product where a combination of different environments may be encountered. A final word on temperature considerations is that all instruments have a limitation on what they can achieve in terms of temperature and humidity in the same step. Although a manufacturer will quote a humidity range and a temperature range for an instrument, it is unlikely that it will be able to achieve the extremes of these ranges at the same time. For example, it is very difficult to achieve 90% RH at very high temperatures so the analyst must check with the manufacturer that the experiment is not beyond the capability of the instrument.

Care should be taken with regards to sample handling due to the tendency of some samples to either lose or absorb moisture from the atmosphere when exposed to ambient conditions. This is most important for experiments where a sample with a specific water content needs to be studied. A common recommendation for hygroscopic materials is to load the sample crucible inside an environmentally controlled enclosure (*e.g.* a glove box) with a dry gas purge. This helps to reduce moisture absorption but is not recommended for samples with any significant water content as it can dry the sample. Ideally, the sample should be removed from its container and transferred to the crucible inside an enclosure of humidity close to that at which the sample is in equilibrium. This, however, is not always practical and can be difficult to predict without running a series of constant humidity experiments. If the sample is to be pre-dried at the start of the experiment then clearly the glove box preparation method is recommended. For extremely hygroscopic materials, it may be necessary to also house the DVS analyser within an environmentally controlled enclosure to avoid moisture sorption during transfer of the sample crucible to the instrument.

A common issue encountered with sample handling is static electricity. Many powders carry a static charge, which can make it difficult to achieve a thin, even layer of the material in the base of the pan and can cause problems with the signal stability. Most instruments have measures to reduce static including grounding of the hang-down wire assembly for the sample and reference crucibles. Copper tweezers can help to dissipate static and are recommended for handling the sample crucibles. The use of quartz crucibles is also recommended for samples where static is an issue.

Sample pan choice is an important consideration. Quartz crucibles (or metal coated quartz) are often favoured due to the lack of reactivity with most materials and the anti-static properties mentioned. These are quite expensive though and are prone to breakage. Cleaning of the crucibles is particularly difficult for stubborn samples and stock of an ample supply is recommended to safeguard against breakages. When loading the sample into the pan, it is particularly important to avoid contaminating the handle as this can stick to the hang-down wire and cause problems with sample unloading. Other pan materials are available including platinum and aluminium. Sealed pans are also available now with the commercialisation of autosamplers with a piercing function. A suitable pan size should also be selected for the sample size required. This should be determined based on the nature of the material (homogeneity of a mixture), the expected behaviour (little mass change will require a larger sample size) and the availability of the material. Typical pan sizes are 20–200 μL; larger pans are available for higher sample masses.

When choosing a sample size for an experiment, the total mass change that may occur during the experiment must be taken into account. Manufacturers state a 'dynamic range' for their systems, which is the maximum mass change possible for accurate recording of results. This is related to the maximum sample mass capacity of the instrument and tends to be in the region of 100 mg for standard instruments. In the case of the higher capacity balances, 500 mg may be stated as the dynamic range.

As with all techniques, the analyst must take care with sample processing steps such as crushing, grinding, milling, *etc.* as this can have a huge impact on the data, especially with respect to the kinetics. These can all affect the particle size distribution of the sample and may even change the physical form (*e.g.* from crystalline to partially amorphous), both of which will affect the rate at which the sample takes on or releases moisture.

4.6 Calibration

Humidity calibration of DVS instruments can be carried out using a number of different approaches. One option is to use deliquescent salts; literature values exist for the humidity at which certain salts will deliquesce at a specified temperature and this can be compared with the observed deliquescence point for that salt inside the instrument. Sodium chloride, for example, will deliquesce at 75.3% RH at 25 °C as referred to previously.[5] The salts used for this calibration must be very pure and should ideally be certified reference materials. The usual approach to this type of calibration is to start the program at a few % RH above the deliquescence point and hold until the rate of water uptake by the sample is approximately linear (*e.g.* one hour). The humidity is then slowly ramped down to a few % RH below the deliquescence point and during this phase, the sample mass will stop increasing. This is the point at which the material is no longer deliquescing and is labelled as the deliquescence point. In order to assign an accurate value to this, the derivative mass signal (dm/dRH) is used; the time point at which the derivative signal is exactly 0 is taken as the deliquescence point, removing any ambiguity from the result. The humidity value at the same time point on the relative humidity signal is then taken as the value for the calibration.

An alternative method is to prepare saturated salt solutions for the calibration. As discussed previously, there are literature values for the relative humidity generated by saturated solutions of some salts at specific temperatures.[5] The saturated solution is placed into the sample chamber and the generated humidity from the instrument is set to a few % RH above the known relative humidity of the saturated salt solution. The generated humidity is slowly ramped to a few % RH below the literature relative humidity value and then back up again. During the increasing and decreasing humidity ramps, a situation will occur where the humidity generated by the instrument is equal to that generated by the saturated salt solution. At this point, the 'sample' mass will stabilise as the sample is in equilibrium with the generated humidity. This is a very similar approach to the one using deliquescent salts and again, the derivative signal (dm/dRH) is used to determine the exact point at which the equilibrium has been reached.

Either of these approaches is appropriate for instruments that have an open loop humidity control system (absence of any humidity feedback sensors in the sample chamber). If the saturated salt solution approach is to be used, the analyst must ensure that a fully saturated solution of the salt has been created or the calibration will be incorrect. Ultra-pure water must also be used in order for the calibration to be as accurate as possible.

For systems containing a closed loop humidity control system, a further option can be considered. Some manufacturers supply verification/calibration standards, which are unsaturated salt solutions of a specific molality (and therefore also a specific water activity). This presents a very straightforward way of calibrating the humidity as the feedback sensors inside the sample chamber directly measure the water activity of the standards used for the calibration.

For any of these approaches, a minimum of two calibration standards should be used (ideally three) within the range of interest (this is usually 5 to 98% RH). The ASTM International standard E2551 contains the full details for calibration of humidity generating thermogravimetric devices (see 'Further Reading' section).

4.7 Applications

Sorption isotherm experiments are often carried out on crystalline materials in order to characterise their equilibrium moisture content or water activity over a specific humidity range as previously discussed. An alternative use of moisture isotherm experiments is to study the behaviour of amorphous materials. They have a very interesting relationship with moisture as they tend to be very hygroscopic and water is known to have a significant effect on their structure and behaviour.[10,11] Water depresses the glass transition temperature and can cause crystallisation to occur rapidly at temperatures where the

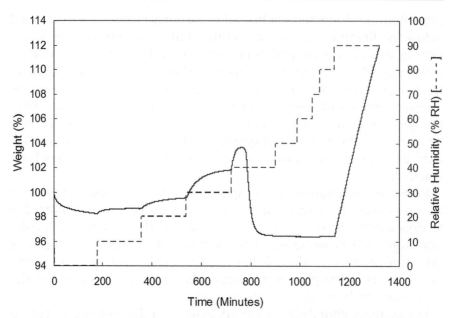

Figure 4.4   Sorption isotherm plot for a sucrose glass at 25 °C. The dotted line represents the humidity program used. The solid line represents the mass change of the sample as a percentage of the initial sample mass (image reproduced from PhD thesis 'A Spectroscopic and Kinetic Investigation into Sugar Glasses', N. E. Hunter, 2009, University of East Anglia).

amorphous form would otherwise be quite stable. This is of critical importance for pharmaceuticals and food products where crystallisation can cause catastrophic failure of the product. Some data from a sorption isotherm experiment for amorphous sucrose is shown in Figure 4.4.

The sample was not pre-dried before the experiment and in the first step, a mass loss can be seen, corresponding to a loss of water that was incorporated into the material during processing (the process does not appear to have reached completion as the mass loss does not plateau before the next humidity step but this did not present a problem for this particular study). During the 40% RH step, the sample reaches a maximum mass and then promptly crystallises, releasing much of the absorbed water as sucrose crystallises to an anhydrous form. Interestingly, this also indicates that in one of the preceding steps, the material must have passed through its glass transition but this is not immediately evident from the curve. During the glass transition, a material changes from a solid-like brittle state to a rubbery, more free-flowing state and as such, there is no mass change specifically associated with it. The large increase in the water uptake at 30% RH though can be taken to be indicative of the sample having passed through its glass transition, becoming more mobile and significantly increasing the rate of water uptake. During the 50 to 80% RH steps, very little mass change occurs as the sample adsorbs a small amount of water onto the surface of the newly formed crystals. In the finally step, the mass increases exponentially due to the sample deliquescing in the high humidity environment. This demonstrates how useful DVS can be in characterising the behaviour of an amorphous material over the entire humidity range.

Constant humidity experiments are useful for a wide range of applications, particularly for acquiring kinetics of vapour uptake by a material at a specific humidity. This may be important for foods, building materials, explosives and pharmaceutics during storage. Using kinetic models, it may be possible to reveal whether a material absorbs the vapour by normal type I diffusion[12] to the core of the structure or whether a band of saturated material forms, which then ingresses into the core (such as the Peleg model for diffusion[13]). This can be important where crystallisation at high water content is a potential issue, for example in pharmaceutical excipients such as trehalose.[14]

4.8 Summary

Understanding how a sample interacts with water and quantifying the amount of water taken up with respect to relative humidity is very important in assessing the stability of numerous materials. DVS is the best analytical technique to make these measurements and is routinely used in numerous industries. While interaction of a sample with water is the usual measurement, many instruments permit the use of organic vapours.

Further Reading

ASTM E2551-13, Standard Test Method for Humidity Calibration (or Conformation) of Humidity Generators for Use with Thermogravimetric Analyzers, ASTM International, West Conshohocken, PA, 2013.

J. W. McBain, The sorption of gases and vapours by solids, 1932.

J. Rouquerol, Adsorption by Powders and Porous Solids; Principles, Methodology and Applications, 1998.

W. E. L. Spiess and W. Wolf, Critical evaluation of methods to determine moisture sorption isotherms. Water activity: theory and applications to food, 1987.

W. Wolf, Sorption isotherms and water activity of food materials: A bibliography, 1985.

References

1. P. Atkins and J. De Paula, *Elements of Physical Chemistry*, Oxford University Press, 2013.
2. T. Lever, P. Haines, J. Rouquerol, E. L. Charsley, P. Van Eckeren and D. J. Burlett, *Pure Appl. Chem.*, 2014, **86**, 545.
3. Q. Zhang, K. Andrew Chan, G. Zhang, T. Gillece, L. Senak, D. J. Moore, R. Mendelsohn and C. R. Flach, *Biopolymers*, 2011, **95**, 607.
4. R. A. Lane and G. Buckton, *Int. J. Pharm.*, 2000, **207**, 49.
5. L. Greenspan, *J. Res. Natl. Bur. Stand.*, 1977, **81**, 89–96.
6. K. S. Sing, *Pure Appl. Chem.*, 1985, **57**, 603.
7. K. Foo and B. Hameed, *Chem. Eng. J.*, 2010, **156**, 2.
8. P. M. Young, H. Chiou, T. Tee, D. Traini, H.-K. Chan, F. Thielmann and D. Burnett, *Drug Dev. Ind. Pharm.*, 2007, **33**, 9.
9. S. E. Hogan and G. Buckton, *Pharm. Res.*, 2001, **18**, 112.
10. Y. Roos and M. Karel, *J. Food Sci.*, 1991, **56**, 38.
11. B. C. Hancock and G. Zografi, *Pharm. Res.*, 1994, **11**, 471.
12. J. Crank, *The Mathematics of Diffusion*, Clarendon Press, Oxford, 2nd edn, 1975.
13. M. Peleg, *J. Food Sci.*, 1988, **53**, 1216–1217.
14. N. E. Hunter, C. S. Frampton, D. Q. Craig and P. S. Belton, *Carbohydr. Res.*, 2010, **345**, 1938.

5 Differential Scanning Calorimetry

Paul Gabbott* and Tim Mann

PETA Solutions Kynance, Kimblewick, Aylesbury, UK
*Email: Paul.Gabbott@btconnect.com

5.1 Introduction and Principles

Energy and its provision is one of the most important aspects of our daily lives, not just for providing light and warmth, but in order to power all of the facilities and devices that we take for granted and use continuously throughout the day. It should be no surprise, therefore, that the measurement of energy and specifically the *flow* of energy into or out of a material is a very significant and important area of analysis; one which covers a wide variety of different materials, ranging from foods and pharmaceuticals, polymers and composites, to clays and minerals. The main instrument used to measure the flow of energy is the differential scanning calorimeter (DSC) and the main measurement that is made is heat flow, normally termed power (P, reported in milliWatts, mW).

Energy usually flows into or out of a material in response to a change in temperature so this is exactly what a DSC is designed to do; it heats or cools a material in a controlled manner and measures the flow of energy into or out of the material whilst it does so. This reveals all of the transitions that occur within the material over the temperature range observed, together with the temperatures at which they occur. The most obvious transitions recorded are melting or

Principles of Thermal Analysis and Calorimetry: 2nd Edition
Edited by Simon Gaisford, Vicky Kett and Peter Haines
© The Royal Society of Chemistry 2016
Published by the Royal Society of Chemistry, www.rsc.org

re-crystallisation, but many smaller, much more sensitive events are also revealed, such as the glass transition (T_g). If appropriate methods are applied then the heat capacity (C_P) of a material can be obtained and reported. DSC is probably the most popular of the thermal analysis techniques if judged by the number of instruments sold and in use. This is due to its wide range of application areas and also the fact that measurements are quick to make. Samples are easily prepared by weighing them into a small crucible (typically an aluminium pan). Scan rates can vary but at a typical rate of 10 or 20 °C min^{-1}, measurements are made in a matter of minutes. Measurements can also be made isothermally over a period of time if a reaction is being investigated or the heat evolved from a biological system is under investigation. This forms the area of isothermal calorimetry, which is discussed in greater detail in Chapter 7 of this book.

5.2 Definition and Nomenclature

5.2.1 The Heat Flow Curve

A DSC measures energy as it flows into or out of a sample as a function of temperature and so the main measurement that is made is referred to as the heat flow curve. The *x*-ordinate can be plotted as either temperature or time. The product of the heat capacity of the sample and the scan rate gives the heat flow:

$$\frac{dq}{dt} = mC_P \frac{dT}{dt} \tag{5.1}$$

A typical example of a heat flow curve is given in Figure 5.1, where the melt of polypropylene is shown as a function of temperature and is observed as a broad endotherm. The peak area has been calculated between user-selected limits to give the *enthalpy* of the transition, in this case the heat of fusion, together with the peak maximum. The term enthalpy is given the symbol H and ΔH is used to describe the energy absorbed by the sample. The run is started at a temperature well below that of the transition so that a flat baseline can be observed before the melt. It will also take a period of time for the instrument to apply and control the heating rate required, a period referred to as the transient, and this must also be allowed for. This appears as a period of instability; often it will look like a step of possibly 30–40 seconds' duration, though this will vary according to the instrument and

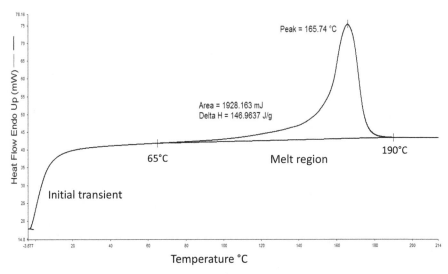

Figure 5.1 DSC trace for the melt of polypropylene, showing analysis for melting point and enthalpy of fusion. The direction of the arrow next to "Endo" indicates the direction of endothermic enthalpy changes.

mass of the sample and sample pan. This is shown in Figure 5.1 though the transient is not relevant to the sample undergoing analysis. The direction in which endothermic and exothermic peaks are displayed can be chosen by the operator, so endotherms can be shown as up or down and it is important to record on the graph which direction has been selected.

When comparing data from different experiments, the y-axis should be plotted as mW m^{-1} g, or W g^{-1} (sometimes termed the normalised heat flow). To some extent, this allows for differences in mass to be taken into account, although best practice would be to use similar masses since other effects such as the broadening of a peak will remain unaffected.

The derivative of the heat flow signal can also be very useful. With DSC, it is used more often with peak analysis to determine whether a given peak is a single event or not. For this, the second derivative is used, which inverts the data, but the shoulders resolve into separate peaks and so can be seen more clearly, such as for the polymorphic transition of phenylbutazone shown in Figure 5.2. The first derivative can be used to help to search for and confirm the position of a T_g since the step corresponding to the T_g will form a peak in the first derivative.

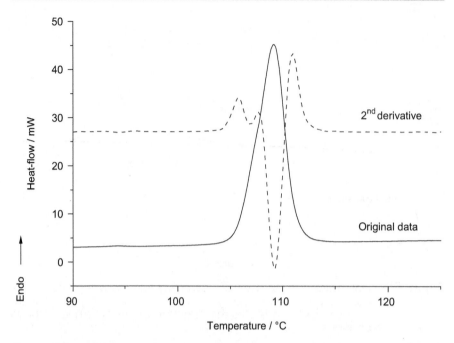

Figure 5.2   DSC trace for phenylbutazone heated at 20 °C min^{-1}. The polymorphic transition is not obvious unless viewed using the second derivative.

5.2.2   Heat Capacity

Though absolute values of C_p are not available from a single heat flow curve, they can be obtained if the contributions from the sample pan, and the empty reference and other possible areas, are taken into account. Essentially, this is done by subtracting a baseline run between isotherms and referencing the height of the remaining curve against a standard material such as sapphire. Standard test methods are usually followed (Table 5.1). The instrument software will usually perform the necessary manipulations automatically, providing the appropriate sample runs have been obtained, and the resulting curve can be displayed in a similar way to the heat flow curve.

The different curves used to produce the heat capacity curve are shown in Figure 5.3. In this example, the baseline has been recorded with empty un-crimped pans run between two isothermal temperatures. A sapphire standard has then been added to the empty sample pan

Table 5.1 A range of ISO and ASTM standard test methods that involve DSC.

Reference	Test method
ASTM D3418	Standard test method for transition temperatures and enthalpies of fusion and crystallization of polymers by DSC
ASTM D3895	Standard test method for oxidative-induction time of polyolefins by DSC
ASTM D4419	Standard test method for measurement of transition temperatures of petroleum waxes by DSC
ASTM D5483 – 05	Standard test method for oxidation induction time of lubricating greases by pressure DSC
ASTM D5885 06	Standard test method for oxidative induction time of polyolefin geosynthetics by high-pressure DSC
ASTM D6186 – 08	Standard test method for oxidation induction time of lubricating oils by pressure DSC
ASTM D6604 – 00	Standard practice for glass transition temperatures of hydrocarbon resins by DSC
ASTM D7426 08	Standard test method for assignment of the DSC procedure for determining T_g of a polymer or an elastomeric compound
ASTM E537 – 12	Standard test method for the thermal stability of chemicals by DSC
ASTM E793 – 06	Standard test method for enthalpies of fusion and crystallization by DSC
ASTM E794 – 06	Standard test method for melting and crystallization temperatures by thermal analysis
ASTM E967 – 08	Standard test method for temperature calibration of DSC and DTA analysers
ASTM E968 – 02	Standard practice for heat flow calibration of DSC
ASTM E1269 – 11	Standard test method for determining specific heat capacity by DSC
ASTM E1356 – 08	Standard test method for assignment of the glass transition temperatures by DSC
ASTM E1858 – 08	Standard test method for determining oxidation induction time of hydrocarbons by DSC
ASTM E2009 – 08	Standard test methods for oxidation onset temperature of hydrocarbons by DSC

Table 5.1 (*Continued*)

Reference	Test method
Standard test method for estimating kinetic parameters by differential scanning calorimeter using the Borchardt and Daniels method.	Standard test method for estimating kinetic parameters by differential scanning calorimeter using the Borchardt and Daniels method.
Standard practice for longevity assessment of firestop materials using differential scanning calorimetry	Standard practice for longevity assessment of firestop materials using differential scanning calorimetry
ASTM E928-08	Standard test method for purity by differential scanning calorimetry
ISO 11357	Consists of the following parts, under the general title of Plastics – DSC
	Part 1: General principles
	Part 2: Determination of the glass transition temperature
	Part 3: Determination of temperature and enthalpy of melting and crystallization
	Part 4: Determination of specific heat capacity
	Part 5: Determination of characteristic reaction-curve temperatures and times, and enthalpy of reaction and degree of conversion
	Part 6: Determination of oxidation induction time (isothermal OIT) and oxidation induction temperature (dynamic OIT)
	Part 7: Determination of crystallization temperatures
ISO 18373-1	Rigid PVC pipes – DSC method – Part 1: Measurement of the processing temperature
ISO 18373-2	Rigid PVC pipes – DSC method – Part 2: Measurement of the enthalpy of fusion of crystallite

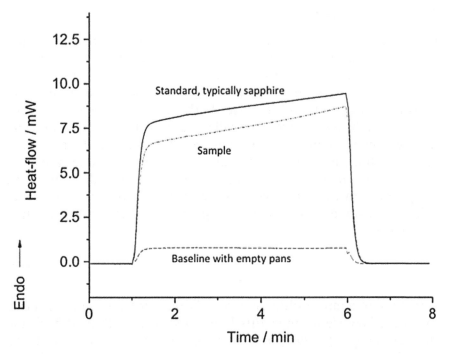

Figure 5.3 Overlay of three DSC traces used to determine the heat capacity of a sample.

and this has then been run under the same method. Finally, after removing the standard, a sample has been run in the same sample pan using the same method. The curves have been presented on the screen, isothermals matched, and the height between the baseline and the sample calculated. The height is then the true heat flow value of the sample itself, which, when divided by the scan rate, gives the heat capacity, from eqn (5.1). The C_p curve may be integrated to give an enthalpy curve that reflects the total energy absorbed by the material.

5.3 Principles of the Technique

All DSC instruments heat the sample and reference with a heater (termed a furnace). The experimental arrangement can have one furnace (heat flux DSC) or two furnaces (power compensation DSC) and the principles of operation are slightly different.

5.3.1 Heat Flux DSC

In heat flux DSC, a single furnace is used to heat both the sample and reference pans, Figure 5.4. Inside the furnace, there are two temperature sensors. The sample is encapsulated in a small crucible and placed on one sensor and a similar empty crucible placed on the other sensor to act as an inert reference. As the furnace is heated, both the sample and the empty reference will initially heat at the same rate and there will be no temperature difference between them. If a transition occurs within the sample, for example if an endothermic transition occurs which demands more energy, then it will fall behind the temperature of the reference and a negative temperature difference will be measured. The heat flow is proportional to the temperature difference:

$$\frac{dq}{dt} \sim \Delta T \tag{5.2}$$

If the temperature difference only is recorded and not converted to a heat flow signal then this is referred to as a DTA (differential thermal analysis) signal, and analysers that do not have heat flux capability, usually older systems, are referred to as DTA analysers.

① Furnace

② Cooling Rods

③ Flange

Figure 5.4 Schematic representation of a heat flux DSC.
Image © TA Instruments. Used with permission.

To convert the ΔT signal to power requires multiplication by an experimentally determined cell constant (k);

$$\frac{\mathrm{d}q}{\mathrm{d}t} = k\Delta T \qquad (5.3)$$

The value of k is usually determined using a certified reference material (CRM). CRMs, often highly pure metals, have defined melting temperatures and enthalpies of fusion and are discussed in more detail in Section 5.6.2.

5.3.2 Power Compensation DSC

In power compensation DSC, separate furnaces are used to heat the sample and reference pans (this was the original DSC design[1]). The power compensations required to maintain the programmed rate of temperature change for the sample and reference are measured and the difference between the signals (ΔP) is plotted. Since the power compensations applied are a direct measure of the energy changes occurring with the sample, the heat flow curve is effectively a read-out of the power compensation circuit. A typical design is shown in Figure 5.5.

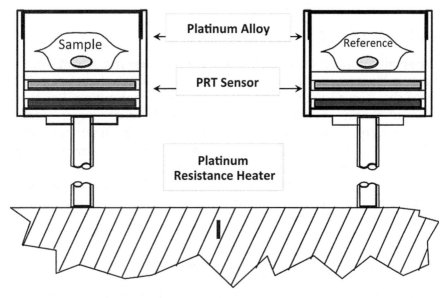

Figure 5.5   Schematic representation of a power compensation DSC. PRT stands for platinum resistance thermometer.

Both heat flux and power compensation DSC systems are often fitted with cooling accessories, normally refrigeration systems, to allow them to operate at sub-ambient temperatures. Fridges can operate to below $-100\,°C$ but to go much below this liquid, nitrogen systems are used. The highest temperature available will depend upon the instrument in use but is normally in the range $450\,°C$ to $700\,°C$, which is fine for organic based materials, though high temperature systems going to $2000\,°C$ or beyond are available for measurement of clays, minerals and other inorganics.

5.4 Instrument Design

5.4.1 Standard DSC

Most heat flux or power compensation DSC analysers require sample sizes of around 1–20 mg, although these may be higher depending upon sample density and pan size. Samples are enclosed in small metal pans and heating rates from 1–$100\,°C\ \text{min}^{-1}$ are typically used.

5.4.2 Micro DSC

The small sample mass used in standard DSC may not be sufficient to allow measurement of some very low energy transitions, especially those that occur in solution such as protein denaturation. For these types of measurements, DSC analysers with much larger sample capacities (typically 1 mL) can be used. They are sometimes called micro DSCs and can hold liquid samples. Typically, slower heating rates $(0.1$–$3\,°C\ \text{min}^{-1})$ are used to ensure the large mass of sample is in thermal equilibrium with the instrument. They may also be operated isothermally to measure changes that occur in a material as a function of time.

5.4.3 Modulated Temperature DSC

Usually, the temperature programme used in DSC is linear. However, it is possible to use non-linear temperature programmes. These have particular benefits to C_p and T_g measurements and discussion of their use is deferred until Chapter 6.

5.4.4 Fast Scanning DSC

Whilst typical scan rates are around $10\,^{\circ}\text{C min}^{-1}$, there are a range of applications that make use of much faster rates. Commercial equipment can currently provide heating rates of up to $750\,^{\circ}\text{C min}^{-1}$. Because power compensation equipment uses separate furnaces, they usually have a smaller thermal mass compared with heat flux instruments and so are most suited to high heating rates. There are two main advantages of fast scanning. The first is increased sensitivity. DSC measures the flow of energy into or out of a material in a fixed amount of time as determined by the scan rate. At slower rates, there is more time available for energy to flow so it flows more slowly, but at fast rates, energy flows more quickly, which means a bigger signal on the Y-axis. So as scan rate is increased, so are the height and sensitivity of a measurement. On a time base, the transition will show as much sharper and higher because the same amount of energy will flow as at slow rates. For example, very small glass transitions can be seen easily at high rates. Secondly, it allows measurement of a sample without giving sufficient time for annealing or recrystallisation events that may occur during slow scanning. So, for example, if looking at a polymorphic material, at slow rates, the structure of the material may change due to melting and subsequent re-crystallisation, whereas at fast scan rates, re-crystallisation can be prevented because it does not have time to occur and so the material analysed will be in its original form.[2] This is particularly helpful when different forms are present or suspected to be present initially. There are a variety of applications to polymers and pharmaceuticals in particular.[3] One disadvantage of using fast rates is reduced resolution so it is important to use helium or a helium gas mix as a purge since this improves heat transfer to and from the sample.

5.4.5 Solid-state (Chip) Calorimeters

A recent innovation has been the development of solid-state (chip) calorimeters. These consist of an electronic chip onto which a very small sample is deposited, usually by melting. Their small mass means they are capable of heating at many thousands of degrees centigrade per minute, allowing the study of scan rate effects well beyond the range of traditional DSC systems. Providing the sample is not destroyed in the heating process, the system can be re-used. However, the technique is limited by the difficulty of accurately weighing the sample size, which produces difficulty in quantifying results, and also by the melting process used to deposit the sample, which destroys the initial crystalline structure so previous thermal history effects cannot be studied.[4]

5.4.6   High Temperature DSC

Standard DSC analysers will normally heat to a maximum of 600–700 °C, but instruments are available offering significantly higher temperatures, with ranges up to 1500–1600 °C and some significantly higher than this. These are referred to as high temperature DSCs. Typically, designs are larger to cope with the higher temperature requirements, some offering combined thermogravimetric analysis (TGA)-DSC measurement (TGA is described in Chapter 3 and TGA-DSC is described in Chapter 10). The principle of measurement used is that of heat flux DSC and the main application area is with inorganic materials including glasses, ceramics, clays and minerals.

5.4.7   UV-DSC

For some applications, it is necessary to irradiate a material with UV light in order to initiate a reaction. With a UV-DSC system, it is possible to irradiate the sample after placing it in the calorimeter and to measure the response by DSC. Such systems are commercially available and sometimes called 'photo DSC' or 'photocalorimetry' systems. The application is largely centred on UV curing of composite materials typically used in the printing and coatings industry and in dentistry. The UV radiation can be transmitted using fibre optic tubing to the DSC furnace, so the UV source can be situated alongside the DSC. Commercially available equipment will offer a choice of sources and filters to allow the selection of the desired wavelength, and experiments are normally conducted isothermally. The time duration for the irradiation can be varied. For some experiments, it is desirable to use a short time period to initiate a reaction, but it may be continuous to develop a full cure. Response times can be very fast depending on the reaction mechanism, so fast data collection rates are needed. These systems allow the investigation of different aspects of cure, to compare and contrast different materials and the ability to develop appropriate compounds.

5.4.8   DSC-Raman and DSC-NIR

In these techniques, Raman or NIR spectra are recorded as a function of temperature (or time) as samples are heated in the DSC. Probes have been developed that allow the spectra to be obtained from re-flected radiation so that as a material is heated, the change in structure can be monitored. The choice of NIR or Raman spectroscopy for this type of work depends on which technique shows the clearer differences in spectra for the different structures that are formed. There are also reports of coupling DSC with FTIR.[5,6] These applications are discussed in Chapter 10.

5.4.9 High Pressure DSC

There are two ways a sample can experience high pressure. If a sample is hermetically sealed in an appropriate sample pan then internal pressure will increase as the sample is heated. Different pan types exist that allow internal pressures up to about 150 bar to be contained, but the actual pressure developed is not measured, and is merely the result of trying to stop volatiles escaping.

Alternatively, external pressure can be applied to an open pan and this is what the term high pressure DSC really applies to. The main use of this is to suppress the loss of volatile components when a sealed pan cannot be used, for example when applying the oxidative induction test (OIT) to oils. In this case, the sample must be open to allow the flow of oxygen to the sample, but pressure is needed to stop the sample volatilising. Commercial accessories exist that allow pressures in the region of 40–50 bar to be applied to the sample. The use of much higher pressures of up to 4–5 kilobars have been reported, but these have been with a modified DSC system.[7]

5.5 Experimental Considerations/Best Practice

5.5.1 Melt Transitions

Many materials are analysed to determine their melting point (T_m) or melting range together with heat of fusion $\Delta_{fus}H$, whether this is simply to ensure consistent quality of a material, such as a thermoplastic polymer, or whether it is part of a more complex investigation, such as determining the crystalline structure of a pharmaceutical or food product. Broadly, there are two types of melt profiles, sharp narrow peaks that result from single crystalline melt and broad, sometimes shallow, melting peaks that result from the varied crystalline structure in materials such as polymers, fats and waxes.

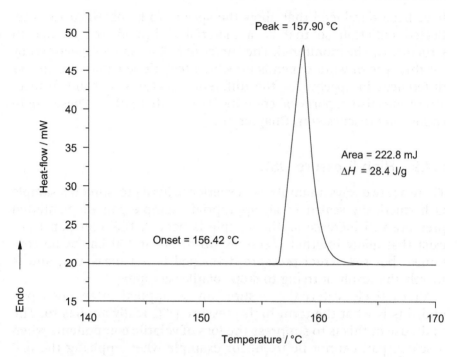

Figure 5.6   DSC trace for the melt of indium at $10\,^{\circ}C\ min^{-1}$.

Melting is an energetic process so, in general, small amounts of material will provide good results.

Pure compounds, such as pharmaceuticals and metals, typify single crystalline materials. Metals are more often run in a DSC in order to calibrate the system or verify that it is working correctly rather than as samples under investigation, but they do show a sharp peak when taken through the melt region. An example using indium is shown in Figure 5.6.

In this example, the data are significantly expanded on the *x*-axis so that the peak shape can be seen. Below the melt temperature, the sample will heat at the pre-programed heating rate as normal, but when the melting point is reached and the crystal structure begins to break down and melt, the sample will hold at the melting temperature and will stay at this temperature until the melting process is complete. But since the DSC will continue to heat, a peak will begin to develop with a straight leading edge as the difference in temperature between the sample and the reference increases. A peak maximum is reached once melting has completed, and the peak will then return to the baseline as the sample temperature returns to that of the furnace. The peak that is formed is not symmetrical, though sometimes it may appear like it, and the melting point is defined by the extrapolated onset to the peak. This is the intersection of two tangents, one drawn from the baseline and the other extrapolated from the leading edge of the peak. The heat of fusion is obtained by integrating the peak area. The peak maximum is not the melting point of a single crystal and it will vary as a function of sample mass and may also vary with sample contact, which will affect the slope of the leading edge. The peak height may also vary with encapsulation technique, so the onset calculation and heat of fusion are the two absolute values obtained from the single crystal melt peak.

Other materials, such as polymers, fats and waxes, may be heterogeneous and contain crystalline structures and so show melting profiles, but unlike pure single crystal materials, these profiles do not show sharp well defined peaks. The melting profile of polypropylene shown earlier in Figure 5.1 is an example of this with a melting range of over 100 °C. This is because there is a range of different crystalline structure within these materials, some regions being better formed and more stable than others and so the materials tend to melt over a broad temperature range. As an example, the peak obtained from the melting of polyethylene may often begin to form just above ambient where weaker crystal structure begins to melt, and the starting point can be difficult to determine (Figure 5.7) and this in turn will affect the area calculated and the heat of fusion value. More stable, better formed crystal structures will melt at higher temperatures and for polyethylene, a peak maximum is normally found above 100 °C.

To try and define an extrapolated onset value for such a broad melting process makes no sense, though the melting range may be meaningful, so the melting point for a material of this type is normally defined as the peak maximum. The precise peak maximum value obtained will vary as a function of the method used, not least due to changes in sample mass, scan rate and encapsulation technique (see Section 5.5.6), so when comparing samples of this type, a common method must be used. As with single crystal melts, the heat of fusion is obtained by integrating the peak area. Most software packages also allow more elaborate calculations, for example the ability to perform partial area calculations between different temperature limits, multiple peak maximum calculations, and other facilities which can be useful when trying to quantify a broad melting transition.

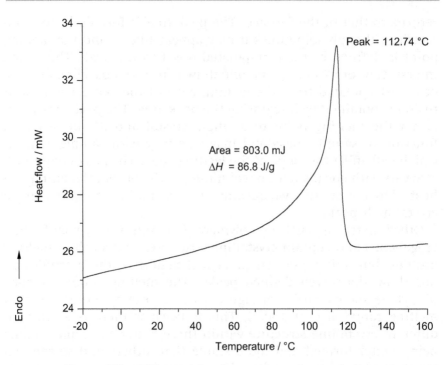

Figure 5.7 DSC trace for the melting of polyethylene.

5.5.2 The Glass Transition, T_g

Whilst melting is observed in crystalline materials as they are heated, a T_g is observed in amorphous materials. The most well-known examples of this are glasses, typically inorganic polymers, but amorphous structure is also found in many organic polymer composites, foods and pharmaceuticals. Many materials that are normally thought of as crystalline may also contain a small amount of amorphous structure, particularly if processed using grinding or milling, which is known to affect crystal structure.

Below the T_g temperature, amorphous materials are hard and rigid, but as they are heated through the T_g region, the molecular mobility increases and they soften and the C_p increases. The heat flow curve reflects the changes that occur in C_p and so it also shows a step change in the T_g region. The C_p change at the T_g is often very small, and so it can be very hard to see, particularly in a composite material, which may be highly filled and cross-linked, so when looking for the T_g, larger sample sizes and scan rates can be helpful since these increase sensitivity. Modulated-temperature DSC is also very useful for T_g measurement (discussed in Chapter 6).

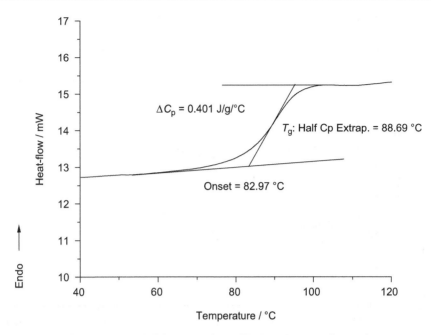

Figure 5.8 A typical DSC curve for a T_g, in this case for polystyrene.

Unlike melting, the T_g does not occur at a fixed thermodynamic temperature, nor is it a fixed point but occurs over a range. It is often defined by a mid-point calculation obtained as the C_p value at half height as shown in Figure 5.8, though the extrapolated onset calculation may be more meaningful as an indication of where the T_g begins. Similarly, the end point calculation or inflection point can be used. For this reason, the type of calculation used should be noted with any values that are quoted.

Quite often, a small relaxation peak can be found overlapping the T_g (Figure 5.9). These peaks are often found if a material has previously relaxed into a lower energy state by slow cooling through the T_g region or following a period of annealing. This heat is known as the enthalpy of recovery and will increase the more a material relaxes. Because it overlaps the step change, it can sometimes make determination of the transition temperature difficult (or even mask it completely). One option using standard DSC is to heat the sample above its T_g then cool rapidly to below T_g. Upon re-heating, it may be possible to see the T_g without the relaxation peak (because the material will not have had time to relax, Figure 5.9). Alternatively, modulated temperature DSC can be used to separate the step change and relaxation signals (discussed in Chapter 6).

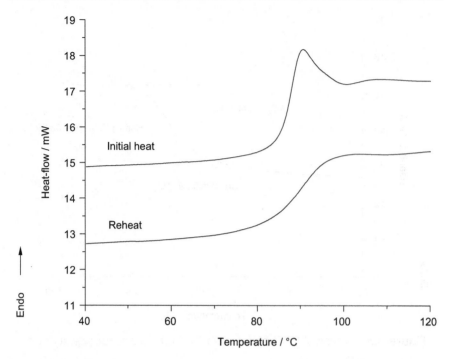

Figure 5.9   The T_g of polystyrene with a relaxation peak superimposed on it (initial heat). Often the relaxation peak will not be observed if the sample is re-heated.

5.5.3   Crystallisation

Crystallisation is an exothermic process. Materials may crystallise during cooling or, depending upon the temperature, during an iso-thermal period after rapid cooling. Sometimes, materials that have been produced in an amorphous state or with amorphous content may crystallise on heating, a process called cold crystallisation. Crystallisation events can also be observed in polymorphic materials when the material re-crystallises into a different form. An example of low density polyethylene (LDPE) crystallising on cooling is shown in Figure 5.10. For a sample like this, it is difficult to quantify accurately the amount of crystalline structure formed by integrating the crys-tallisation peak since the end point of the process cannot be easily determined, so measurement of the extent of crystallinity is best done by measuring the heat of fusion on the re-heat and comparing against a known reference curve. The onset of crystallisation may also be delayed by super cooling (under-cooling), which can occur with any sample, and, at slow cooling rates, crystallisation may occur at lower

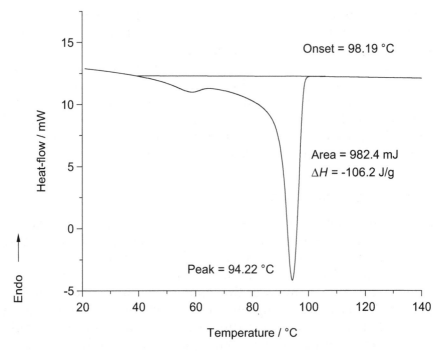

Figure 5.10 Crystallisation of LDPE on cooling by DSC.

temperatures than with fast cooling rates. It is for this reason that DSC instruments are normally calibrated for temperature upon heating.

5.5.4 Reactions

Reactions can also be observed by DSC and will appear as an exothermic peak. Cure reactions in composites are a typical example. Some reactions may occur quite quickly especially if initiated by UV light but many reactions will occur over prolonged periods of time, so slow scan rates or isothermal measurements (Chapter 7) may be preferred. If the reaction is initiated by heating to a higher temperature and is to be measured isothermally, then the sample should be heated rapidly from low temperatures to minimise the extent of cure before the isothermal temperature is reached. Often it is best to run a temperature scan first to determine the best isothermal temperature to use, so that the reaction does proceed in a reasonable time period. Note that it will be important to examine the data carefully to see any peaks that are generated. Sometimes, the transient may interfere with the measurement if it occurs rapidly, in which case, this should be subtracted after re-running the fully reacted sample to produce a curve for subtraction. The heat of reaction can be obtained by integrating the area under the peak.

When looking at large, possibly fast, reactions, the furnace of the DSC may overheat before cooling back to the programmed temperature. If this happens, the heat flow curve when shown against temperature may loop back on itself. The data will look better plotted *versus* time, but it is better to repeat the measurement with a lower sample mass for accurate enthalpy determination.

5.5.5   Loss of Volatiles

Volatile loss can be a very significant event since it can take a large amount of energy to volatilise a material, and hence very large endothermic peaks can be produced. Water is a commonly encountered volatile material and if water-containing materials are run, the loss of moisture over the temperature range above ambient can totally mask everything else that occurs. It is for this reason that such materials are best run in sealed pans capable of withstanding the internal pressure build up so that volatile loss can be minimised.

If the volatile loss itself is to be measured then it may be worth considering a pan with a very small hole in the lid. These are available from most manufacturers or a pin can be used to create the hole. The hole is typically so small that very little material escapes below the boiling point of a material, but then as the partial pressure increases above atmospheric, the material will escape quickly giving a sharp peak. This also improves the resolution of peaks from solvates or hydrates where the volatile material may be lost above its boiling point.

Figure 5.11 shows water of hydration lost from copper sulfate pentahydrate scanned at 5 °C min^{-1} in a sealed pan with a 50 μm hole in the lid. There should be three main peaks from the hydrate loss in the temperature range to 300 °C. Two molecules of water are lost around 100 °C, giving the initial doublet, two more at a slightly higher temperature and the final molecule above 200 °C. The exact peak shape and temperatures recorded will depend upon the method of encapsulation, which determines how easily the volatiles are lost from the pan, along with scan rate. The small hole in the lid sharpens peak shapes but can also shift the loss of volatiles to higher temperatures. Since volatile loss appears as an endothermic peak, it can be confused with melting so care needs to be taken with interpretation, and it may

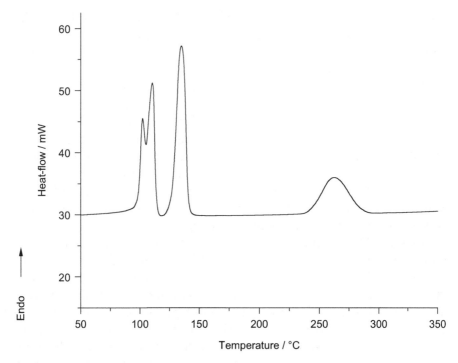

Figure 5.11 DSC trace for loss of water of hydration from copper sulfate pentahydrate heated at 5 °C min⁻¹ in a sealed pan with a 50 μm hole in the lid.

be helpful to run a TGA curve for comparison. To dry a sample will potentially change its properties and if the sample is to be measured as received, the choice is either to encapsulate in a sealed pan to prevent volatile loss or run at a very high scan rate. On the other hand, if the moisture loss is not large and does not significantly affect sample properties, then it is not necessary to encapsulate in a sealed pan. A small broad peak may often be found from moisture loss in this situation.

5.5.6 Pan Selection

Samples are not put directly into a DSC furnace but are enclosed in a suitable sample pan (sometimes called a crucible) so that the DSC furnace stays clean and free from contamination. The most common material for a pan is aluminium, but all manufacturers offer a range of pans for different applications and a range of sealing devices for enclosing them. It is important to make sure that the sample cannot form an alloy with the pan material and the pan must be fit for the temperature range being used, bearing in mind that aluminium should not be heated above 600 °C where it will soften and begin to alloy with the furnace.

Aluminium is a fairly soft metal and can be cold-welded to form a seal so allowing the use of hermetically sealed pans. One potential issue with this is that aluminium can deform under internal pressure, and this must to be avoided during measurement. In fact, most samples will give rise to a large increase in internal pressure if sealed and so aluminium pans are normally vented, either by using a lid with a hole or by piercing the lid first to make sure it has a small pinhole. Some encapsulation systems crimp but do not hermetically seal and so these do not need any further precautions.

Most solids will contain a small amount of moisture, which may add to the pressure build up in a sealed pan, but with a vented pan, moisture will escape giving a small broad endotherm in the 50 to 80 °C temperature range depending on parameters such as scan rate and size of hole. As the amount of moisture increases, this can become an issue because it may mask other transitions in this region. This is particularly true of amorphous freeze-dried materials when the T_g is to be measured and where the moisture should normally be retained. Significant volatiles can also be produced from condensation reactions, for example when curing phenolics, and these need to be encapsulated in sealed pans, preferably steel, to hold the internal pressure. Other materials such as water containing biological materials produce significant volatiles during the course of the scan. Steel pans capable of holding the pressure may be best for all these applications, although for freeze dried materials, a hermetically sealed aluminium pan will usually be acceptable since the maximum temperature will not normally be very high or the water content significant. It is always best to check the manufacturer's recommendations.

High pressure pans retaining about 150 bar internal pressure are also available. One of the main uses is for hazard analysis in which case the pan is gold coated to prevent unwanted catalysis by other materials, but can be used for other systems where high pressure build up is expected.

In all cases, pans should be clean and kept in closed containers before use, and discarded if misshapen or dirty. Pans should not be overfilled as there is the potential for material to overflow onto the furnace if the sample expands in volume with temperature or on melting.

5.5.7 Sample Preparation and Encapsulation

The sample should be in good thermal contact with the pan so that transitions are sharp and the sample will not collapse and cause spurious peaks as it heats. If a sample is poorly encapsulated then heat transfer will be slow, allowing the build-up of thermal gradients across the sample. This will broaden transitions and resolution will be poor. If using a powder, this should be compacted into the base of the pan as much as possible; or select an encapsulation process that compresses the sample. It may be possible to use a pestle and mortar to grind larger pieces, though not if the sample is prone to poly-morphism, in which case a sieve can be used to isolate a fine particle fraction. If a polymer granule or composite is to be measured, a flat base should be cut into the sample so it can sit at the base of the pan. It is useful to obtain a range of cutting instruments for this purpose. The top of the sample does not have to be flat. The pan should be checked after encapsulation to make sure the base is flat and that no particulate contaminants are on the outside surfaces.

5.5.8 Sample Mass

Large sample sizes increase sensitivity but lead to higher thermal gradients across the sample and so reduce resolution. There are also more difficulties getting a large sample in a pan, there is more po-tential for resulting contamination, and there may be issues with collapse as the sample softens and melts. Small sample sizes improve resolution and are generally easier to work with but will reduce sen-sitivity. One of the most common mistakes is to use too much sample, which increases the risk of instrument contamination from sample escaping from the pan during measurement.

Melting enthalpies are generally large so 3 to 5 mg of sample may be sufficient to enable accurate determination of these transitions. With powders of low density, typical of some pharmaceuticals, lower masses are often used since the pan may not be big enough to hold more. If polymer samples are to be examined, then these are often of higher density and masses of around 10 mg may be better since these exhibit broader melts and may need the increased sensitivity.

If a low energy transition such as a small T_g is to be measured then it can be useful to maximise the sample size in order to maximise sensitivity, providing that the pan can still be properly encapsulated, so pans must not be overfilled. Note, however, that if a large sample size is used in order to see the T_g, this may not be appropriate for subsequent melting. Errors can occur during the melting of a large sample because of collapse and possible leakage out of the pan, resulting in erroneous results and instrument contamination, so do not melt large sample sizes. If melt information is also required, repeat the run with a smaller sample size appropriate to melting, and do not try to obtain all the information from just one run.

5.5.9 Temperature Range

The initial temperature should be below that of the first transition of the sample to allow time for the transient to complete and for a period of straight baseline to be produced. This may mean using a sub-ambient temperature and so a cooling accessory will be needed. Note that the sample should be loaded at room temperature to ensure no condensation of water from the atmosphere within the DSC chamber.

The final temperature should be high enough that the sample has progressed through its phase transitions, but not so high that decomposition occurs (many samples, especially pharmaceuticals, degrade on melting), which is usually observed as a noisy baseline above the melt. If the decomposition temperature is not known then it is good practice to run the sample in a TGA first to determine where decomposition begins, and then keep within the stability range of the sample. If this is not possible then the experiment should be watched in real-time and stopped manually if degradation is suspected. Faster heating rates may also reduce the extent of degradation. Sometimes, volatile loss can cause an endothermic response, but if the signal begins to drift and becomes noisy, the experiment should be aborted.

5.5.10 Scan Rate

With standard DSC equipment, the most common scan rates are 10 and $20\,^{\circ}\text{C}\ \text{min}^{-1}$. Generally, the faster the scan rate the higher the sensitivity that will be achieved. However, thermal gradients will also increase with faster scan rates, reducing resolution (similar to the effect of increasing sample mass). Kinetic effects also need to be considered; fast scan rates may not give time for some events to occur and can inhibit some reaction processes altogether. Making measurements at two scan rates, typically an order of magnitude apart, can also be very useful, as true thermodynamic events, such as melts, will appear at the same temperature while kinetic events, such as crystallisations, will appear at higher temperatures with increasing heating rates. When very high scan rates are used, it is important to make sure the data capture rate of the instrument is fast enough that resolution in the data is not lost.

5.5.11 Artefacts

Sometimes, transitions are unexpectedly small or occur at an unexpected temperature or may have an unexpected shape and the question arises as to whether they are really attributable to the sample. Artefacts can occur for a variety of reasons ranging from sample or furnace contamination to pan movement due to vibration or the instrument being knocked. Artefacts can also result from a failure or fluctuation in services to the instrument such as gas or power supply or from faults in the instrument itself, or it may be because the sample was not encapsulated well. Interpretation starts with understanding the sample and knowing what transitions might reasonably be expected to occur. It is also good practice to repeat the measurement and ensure suspect events are reproducible, although contamination in the system may itself cause a repeatable event so the system should be clean. If an instrument fault is suspected, the run should be repeated without any pans to see how the baseline appears. Events that are extremely sharp are also suspect since most thermal transitions occur over a range. Events that may cause sharp peaks include sample movement in the pan and explosive rupture of the pan seal. If enthalpy values are not as expected, check that the sample mass has been measured accurately and entered into the system correctly.

5.5.12 Purge Gas

All analysers use a purge gas to control the environment around a sample, to remove volatiles, and to prevent condensation in sub-ambient systems. The most common purge gas is nitrogen since it inhibits oxidation of samples and is low in cost. Air and oxygen can be used within the temperature limits specified for your instrument. Argon can also be used though this has poor heat transfer characteristics, resulting in poorer resolution and general performance although it can work well at higher temperatures. Helium is used as a purge when liquid nitrogen cryogenic systems are used since nitrogen will condense at the low temperatures reached, and it is also used when very fast scanning rates are chosen since it improves heat transfer and thus resolution. It is rare that other gases are considered but if a reducing atmosphere is required, gases such as hydrogen or

ammonia should only be used if the instrument specifications permit
it and within sensible concentration limits.

5.6   Calibration and Verification

5.6.1   Verification

Verification is the process of checking and confirming that an instrument is performing within acceptable limits. Often, indium is used for
this purpose since its melting point of 156.6 °C is within the range for
many of the measurements made by DSC and the sample can be re-
used provided it is not heated above 180 °C. Indium is soft and should
be flattened before being placed in the pan. Variations of more than a
degree can be found if the indium sample does not have good thermal
contact. The frequently suggested method of pre-melting does not
really help since indium does not flow when it melts. The other melting
point standards commonly used over the lower temperature range are
lead, tin and zinc. These must be used only once and then discarded
since the melt will tend to alloy with aluminium. To verify that a system
is working properly, a standard material should be heated through the
melting region under the conditions required for subsequent samples
and the melting point obtained from the extrapolated onset, and heat
of fusion values, noted. If these fall within acceptable limits, then work
can proceed. If not, then the system will need to be re-calibrated. Limits
such as $+/-$ 1 °C of the melting point and $+/-$ 1% of the value of
heat of fusion of indium could be chosen, though these limits do not
suit every application or industry and thought should be given to this
when setting up standard protocols.

Verification should be performed regularly. In some industries, this
may mean once a day (typical in the pharmaceutical industry) but
other industries and applications under less strict regulation normally accept longer time intervals, and again thought should be given
to how often this needs to be done.

5.6.2   Calibration

Calibration should be performed if verification has failed, or if a new
set of conditions are to be used. It is pointless to schedule calibration
on a regular basis since it may not be necessary, and in any case,
verification should be performed immediately after calibration to
make sure calibration has been done properly.

The process of calibration is generally fairly quick and easy to perform following the manufacturer's instructions for the software and analyser in use. Numerous melting point standards for DSC exist, most of which are available as certified reference materials,[8] listed in Table 5.2. When choosing a standard, care should be taken with encapsulation to make sure the standard used does not react with the pan material when it melts. In many cases, standards should be used once and discarded, possible alloying between a standard and the pan is one reason for this. A range of reference materials can be obtained from instrument manufacturers and also from LGC Standards. Calibration is normally carried out on heating since super cooling effects mean that the crystallisation point of materials may not be accurate. This does not mean that no attempts have been made and there are references to the use of liquid crystalline and other materials for this purpose[9,10] but commonly, DSC systems are not calibrated for cooling.

CRMs that cover the whole temperature range of use are not needed, since the response of most instruments is linear. Two standards with reasonably different melting temperatures should be sufficient (note that indium and tin are quite close and should not be used without a third standard). All experiments should be run under an inert atmosphere to prevent oxidation. Sub ambient performance should be considered especially if a liquid nitrogen cooling system is in use in which case some of the organic materials shown in Table 5.2 may be used. They will not be available as CRMs but they are better than nothing. In general, mercury is to be avoided, and care may be needed with other standards to make sure there is no interaction with the pan material.

The purpose of calibration is largely to adjust and correct for any inaccuracies in the temperature or energy measurements of an analyser, and in the case of heat flux instruments, to determine the proportionality constant. It is understandable that different thermocouples or other temperature sensors could vary in performance and this needs to be corrected, but temperature calibration also corrects for any thermal lag that exists in a system. Thermal lags exist because it takes time for the energy of the DSC furnace to penetrate the sample, and since the temperature sensor is always separate from the sample, it will measure a potentially different value. The extent of any thermal lags will be scan rate dependent, so this means that calibration must be performed for each scan rate of interest, and also for any other fundamental change of method such as change of cooling system or change of purge gas, though air, nitrogen and oxygen all

Table 5.2 Reference materials commonly used for verification and calibration of DSC analysers. The values quoted in this table were obtained on-line and should be used as a guide. When materials are purchased, the certified values accompanying them should be used for calibration purposes.

Standard	Melting point or transition temperature (°C)	Heat of fusion (kJ mol^{-1})
Cyclopentane	− 151.16	Crystal rearrangement
Cyclopentane	− 135.06	Crystal rearrangement
Mercury	− 38.8	11.47
Cyclohexane	6.5	n/a
Gallium	29.8	5.59
Benzil	95	23.26
Indium	156.6	28.42
Tin	231.9	7.03
Lead	327.5	4.76
Zinc	419.5	108.26
Bismuth	544.5	53.0
Aluminium	660.0	10.83
Silver	961.8	11.28
Gold	1064.0	12.55
Platinum	1554.9	16.74

have similar thermal properties so these can be switched without issue.

Calibration should be carried out under the conditions of use, which means the same scan rate that will be used to collect data. If different scan rates are required, it may be necessary to save different calibration files for each scan rate. Typically, there is little variation in thermal response between different types of aluminium pans so it is not necessary to recalibrate if changing from one to another, but it is prudent to verify this for your own system. Recalibration will be required if pans made of other materials are to be used.

5.7 Applications

5.7.1 Polymers

Polymeric materials are widely investigated by thermal methods whether to ensure quality, to develop compounds and mixtures with specific properties or for general troubleshooting and problem solving.[11-15] This section will deal with thermoplastic materials that melt or flow when heated, while the section on composites will deal with thermoset materials.

One of the most common uses of DSC is to check the quality of a material by melting. This has become more significant with the increasing use of recycled material, which varies in composition. It is quite straightforward to weigh a small amount of sample into a pan and to scan over the desired temperature range (for instance, the melt of polypropylene shown earlier in Figure 5.1).

However, for a thermoplastic, one of the first things to consider is what is termed the 'thermal history' of the material, which refers to the thermal conditions the sample experienced before analysis. These may affect the crystal structure and consequently the melting profile. Much information can therefore be gained by cooling a sample after melting in the DSC and reheating. Cooling under defined conditions will impart a known and reproducible thermal history to the sample. The effects of thermal history on poly(ethylene terephthalate) (PET) are shown in Figure 5.12. The initial heat is of the as-received material and shows a T_g with a significant relaxation peak, followed by a

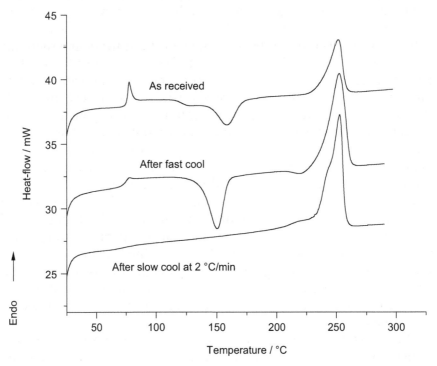

Figure 5.12 DSC traces for various PET samples with different thermal histories.

recrystallisation process showing a doublet before the final melt. There are clear differences in the structure following fast cooling and following slow cooling. In general, the initial heat gives information about what happened to the material before it was received, but the reheat gives more information about the underlying properties of the material. For example, if a product from a production line is causing concern, then both the initial heats and reheats should be compared. If the reheats are different then this implies that the material used in production is different and the raw materials should be checked. However, if the reheats are the same but the initial heats vary then this suggests an issue with the production process.

Comparison of different materials is therefore a key approach to gaining information from DSC. As mentioned in the section on melting, data to be compared should be normalised (presented on a mW g^{-1} scale) to allow better comparison. It may be possible to obtain reference data for typical polymer melts, but because of the many grades and types of any given polymer, it is probably best to build up a library of curves from actual samples. Identification from the melting point is possible, but since the melting point may not be unique to a given polymer, identification of a completely unknown material may not be possible, but it will help to narrow down the options.

If the extent of crystallinity of a polymer is to be measured then this is normally obtained by comparing the heat of fusion of the melting peak with a sample of known crystallinity. It can be interesting to note differences in crystallisation characteristics. Sometimes, materials that have similar melting profiles can be distinguished by their crystallisation behaviour, so isothermal crystallisation tests can be quite helpful. This involves cooling rapidly to an isothermal temperature, where crystallisation will occur over a reasonable time frame, and comparing results for different samples.

The T_g temperature of polymers can also be used for identification and characterisation. Many factors influence the T_g both in terms of the actual measured temperature and the temperature range over which it occurs, and these can all be examined using DSC. These include the effects of changing molecular weight, the use of fillers and additives including plasticisers and stabilisers, the effects of mixing or co-polymerisation with other materials and the effects of ageing and annealing.

The T_g itself can be a small event and since the height of the transition is proportional to the amount of amorphous material present, it will be even smaller if the material is partially crystalline or highly filled when the material of interest forms only a small proportion of the sample. This is evident in the trace of PET in Figure 5.12 where, after slow cooling, the T_g is very hard to see because the material is at this point substantially crystalline. Larger sample sizes may be needed to help improve sensitivity, though excess amounts that may leak out of a pan or cause collapse and other issues should be avoided. Modulated or fast scan techniques are useful if the transition is very hard to see, and generally, it is helpful to use faster rates to improve sensitivity.

Another frequently employed test is the oxidation induction time (OIT) test. Manufacturing processes frequently require a polymer to be heated well above melt temperature in order for it to flow and be manufactured into the required product. It is important that the polymer is stable during this process and the OIT test is a way of establishing this. The material is placed in an open pan (or a pan with a suitably large hole) so that the sample can interact with the purge gas. It is heated to a specified isothermal temperature under an inert atmosphere, usually nitrogen, and once stabilised, the purge gas is switched to oxygen and the time taken to oxidative decomposition is measured. This is the oxidative induction time and, depending on the polymer, this will vary with the amount of stabiliser or other additives present. See standard methods in Table 5.1.

5.7.2 Foods

Many foods are polymeric, so the comments made about polymer applications also apply to a range of foods. Investigations can be made into a wide range of food materials and ingredients including fats and butters, starch, sugars, frozen systems, proteins and other areas. This section will deal with the main applications that are used, those of fats and butters of which chocolate is very important, sugars and frozen materials.

Figure 5.13 shows the trace for a sample of chocolate that has been heated, cooled and reheated. The crystalline structure revealed in the initial heat is quite different to that of the re-heat. Chocolate can exist in completely different crystal forms; it is polymorphic in nature, so in manufacturing, the chocolate has to be carefully processed to produce the required crystal structure. Melting profiles can be quite complex so partial area calculations or percent crystallinity curves are also very useful in defining melting profiles.

Frozen sugar solutions are commonly used in foods such as ice cream, or in the manufacture of freeze-dried products, and these show interesting thermal behaviour. When a sugar solution is cooled

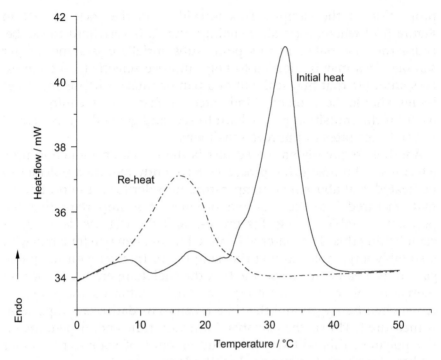

Figure 5.13 Comparison of the heat and reheat DSC traces for a sample of chocolate.

below freezing point, water will begin to freeze out as a pure ice phase and the remaining unfrozen solution will increase in concentration. This process does not continue indefinitely to give pure ice and pure sugar, but instead, a temperature is reached where the remaining solution forms a glass. The T_g of this glass is dependent on the amount of water present and in a freeze-dryer, the concentration of water can be lowered so that an amorphous material with a higher T_g can be produced, which is therefore stable at higher temperatures.[16] This process is used to stabilise materials in both the food and pharmaceutical industries. An example of the glass transition region of a frozen sucrose solution heated in a DSC is shown in Figure 5.14. There are other events shown in this trace and these have been explained in detail elsewhere.[17]

5.7.3 Composites, Adhesives and Curing Systems

The drive in recent years in markets such as aerospace and advanced engineering materials is to move away from metals and alloys and towards composites. These rely on a mixture of a resin with a matrix,

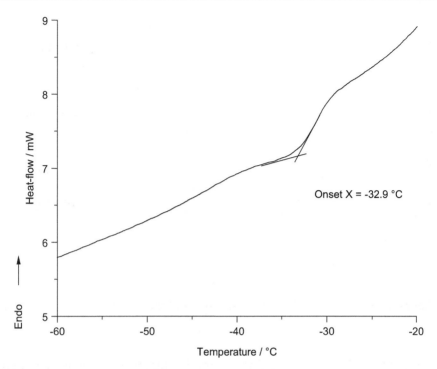

Figure 5.14 DSC trace of a frozen 15% (w/v) sucrose solution heated at 10 °C min^{-1} showing the glass transition region.

the matrix being a material such as glass, carbon or aramid fibre. The benefit of using composites is that they can have great strength but are considerably lighter than the alloy components that they are intended to replace.

It is important to measure the T_g of a composite since it will increase as a function of cure and so give an indication as to whether the cure has been properly performed. Historically, dynamic mechanical analysis (DMA) has been used to measure the T_g of cured systems because the T_g is often difficult to measure by DSC, particularly where the resin is very heavily filled and cross-linked, and as a consequence, the change in energy associated with the T_g is very small. The use of fast heating rates in DSC has allowed the measurement of these T_gs by increasing the sensitivity of the measurement by orders of magnitude, shown with poly(methyl methacrylate) (PMMA) in Figure 5.15. One further advantage of using these accelerated rates is that any post cure of the resin that occurs straight after the T_g can be delayed to a higher temperature due to the slower kinetics of the curing reaction itself so the transition is clearer. In addition, if the

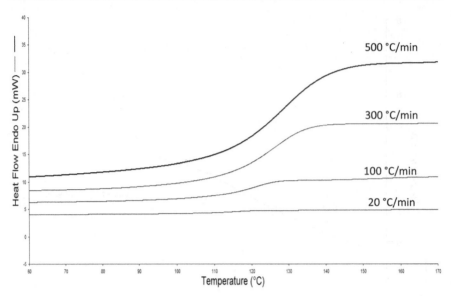

Figure 5.15 The T_g of PMMA at heating rates from 20 °C min^{-1} to 500 °C min^{-1}.

sample has picked up any plasticisers during storage, *e.g.* water, then these may not be lost from the sample during the scan, depending upon the temperatures involved, so the plasticised T_g rather than a dried or partially dried T_g may be measured. Loss of moisture is also obvious if it occurs. The use of modulated temperature DSC for T_g measurement is discussed in Chapter 6.

DSC is the only technique that can measure the energy associated with the cure reaction and, as a consequence, DSC is commonly used in the analysis of resin systems, the composite material before curing (prepregs) and the final cured composite, together with the adhesives used to bond them together. Frequently, a quality test of the resin and prepreg is run before use in manufacturing. In this case, the resin should start to react at a pre-defined temperature and have a total area of reaction within a set range. If the sample does not meet these parameters then it will often be discarded as it is likely to have already partially reacted in storage/transit. The same would apply to an adhesive that is to be used to bond composites together. The finished product itself will also be analysed and any residual resin that is unreacted will continue to react and will be identified as an exotherm above the T_g.

DSC can also be used to give a kinetic analysis of the cure process. Calculations usually based on Arrhenius kinetics can be used to analyse the cure reaction as a function of temperature or from a series of isothermal scans performed at different temperatures. This provides values for the activation energy, order of reaction and rate constant, which is the basis for subsequent predictive calculations of performance outside the measured limits, see for example ASTM methods E2923 and E2041 in Table 5.1. Software to perform these calculations is provided by most manufacturers.

5.7.4 Pharmaceuticals

Knowing the polymorphic composition of pharmaceuticals is importance, since distinct forms may exhibit quite different properties, some of which may be undesirable. Crystal forms are most easily identified by melting point. Since different forms may have similar melting points, it is common practice to use a range of scan rates when looking at polymorphism, typically 1–10 °C min^{-1}, so that the exothermic crystallisation events may be seen between closely spaced melting events. It is particularly important to look for exothermic crystallisation peaks following melting, as this indicates metastable forms were present in the sample. The effect of using two scan rates is highlighted in Figure 5.16. It is clear from the slower scan rate that interconversion is occurring.

Where a pharmaceutical is fully or partially amorphous, the height of the T_g can be used to determine the degree of amorphous content. Fast scan rates are useful when the T_g is very small to increase the size of the event and so increase the accuracy of measurement.

A classic use of DSC is for estimation of purity. The principle is based on the Van't Hoff law of freezing point depression and is generally included in the analysis software provided with the instrument. An impurity will lead to a change in shape of the leading edge of the melt peak of the major component and it is analysis of this leading edge that is used to calculate the impurity. As a consequence, the calculation is carried out on a small (often approximately 1 mg) encapsulated sample at a slow scan rate of around 1 °C min^{-1}. The technique only works for impurities that form an eutectic melt with the main component, so if the impurity shows a distinct separate melt, then this type of purity analysis will not be suitable. This method cannot be used on samples that are less than approximately 97% pure as the corrections applied become too great to give a reliable result.

In formulated medicines, it is important that there are no undesirable drug-excipient interactions that may result in a loss of product performance. DSC analysis offers a quick screening tool for this

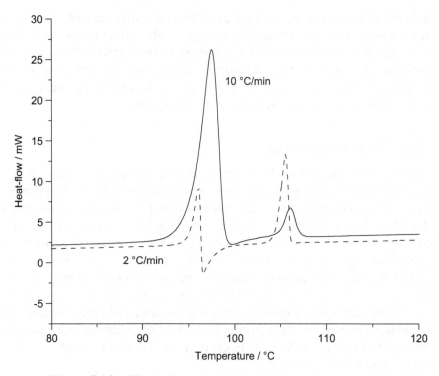

Figure 5.16   Effect of scan rate on polymorph conversion.

purpose, whereby DSC scans of the individual components are compared with data from a scan of the drug-excipient(s). If obvious changes
are seen, for example changes in the enthalpy of the melt peaks or the
appearance or disappearance of peaks, then this suggests that there is a
physical or chemical interaction in the mixture. While physical interactions such as one material dissolving in another as it is heated do not
necessarily indicate that there is an issue with interaction at lower
temperatures, a lack of any changes is generally taken as an indication
that there are no significant interactions. Note that this does not
guarantee that there are no interactions, and other more sensitive approaches such as characterisation of a larger sample size by isothermal
calorimetry or classical HPLC should always be used to check.

5.8   Summary

DSC is probably the most widely used thermal analysis technique,
because of its ease of use, wide range of applications and ability to
measure almost any phase transition in solid (and in some cases

liquid) samples. Careful selection and variation of experimental parameters can be used as a powerful aid to data interpretation. Comparison with data obtained from other techniques, particularly TGA, also gives fundamental insight into the nature of any transitions occurring. Discussion of modulated temperature DSC is covered in Chapter 6 and sample-controlled DSC is covered in Chapter 11.

References

1. E. O'Neill and E. Watson, *U.S. Pat* 3263484.
2. C. McGregor, M. H. Saunders, G. Buckton and R. D. Saklatvala, *Thermochim. Acta*, 2004, **417**, 231.
3. J. L. Ford and T. E. Mann, *Adv. Drug Delivery Rev.*, 2012, **64**, 422.
4. W. Chen, D. Zhou, G. Xue and C. Schick, *Front. Chem. China*, 2009, **4**, 229.
5. B. Degamber, D. Winter, J. Tetlow, M. Teagle and G. F. Fernando, *J. Meas. Sci. Technol.*, 2004, **15**, L5.
6. S. Y. Lin and S. L. Wang, *Adv. Drug Delivery Rev.*, 2012, **64**, 461.
7. J. Ledru, C. T. Imrie, J. M. Hutchinson and G. W. H. Höhne, *Thermochim. Acta*, 2006, **446**, 66.
8. G. Della Gatta, M. J. Richardson, S. M. Sarge and S. Stolen, *Pure Appl. Chem.*, 2006, **78**, 1455.
9. ASTM E967 – 08 (2014) *Standard Test Method for Temperature Calibration of Differential Scanning Calorimeters and Differential Thermal Analyzers*.
10. J. D. Menczel, *J. Therm. Anal.*, 1997, **49**, 193.
11. V. B. F. Mathot, *Thermal Analysis of Polymers*, Hanser, 1993.
12. G. W. Ehrenstein, G. Riedel and P. Trawiel. *Thermal Analysis of Plastics, Theory and Practice*, Hanser, 2004.
13. J. A. Bevis and M. J. Forrest, in *Principles and Applications of Thermal Analysis*, ed. P. Gabbott, Blackwell, 2008, p. 164.
14. J. D. Menczel and R. B. Prime, *Thermal Analysis of Polymers: Fundamentals and Application*, Wiley, 2009.
15. B. Wunderlich, *Thermal Analysis of Polymeric Materials*, Springer, 2005.
16. F. Franks and T. Aufrett, *Freeze-drying of Pharmaceuticals and Biopharmaceuticals: Principles and Practice*, RSC Publishing, 2007.
17. M. J. Izzard, S. Ablett, P. J. Lillford, V. L. Hill and I. F. Groves, *J. Therm. Anal.*, 1996, **475**, 1407.

6 Modulated Temperature Differential Scanning Calorimetry

Vicky Kett

School of Pharmacy, Queens University Belfast, 97 Lisburn Road,
Belfast BT9 7BL, Northern Ireland, UK
Email: v.kett@qub.ac.uk

6.1 Introduction and Principles

Modulated temperature differential scanning calorimetry was introduced by Reading and is a software modification of the standard DSC technique.[1,2] The instrumentation is, therefore, exactly as described in Chapter 5.

The experimental parameters that must be set are heating rate (as in conventional DSC) but also the amplitude (ranging from ± 0.01 to $10\,^{\circ}C$) and period (ranging from 10 to 100 s) of the modulation. The user should also pay consideration to the choice of pan type, sample size and preparation, gas type and flow rate.

6.2 Definition and Nomenclature

The ICTAC approved name for the technique is modulated temperature differential scanning calorimetry, which is abbreviated to MT-DSC.[3] When MT-DSC was first introduced, there was some controversy relating to the name of the technique and of the additional

Principles of Thermal Analysis and Calorimetry: 2nd Edition
Edited by Simon Gaisford, Vicky Kett and Peter Haines
© The Royal Society of Chemistry 2016
Published by the Royal Society of Chemistry, www.rsc.org

signals obtained when modulated methods were employed. In early papers, reference was made to alternating DSC (ADSC), temperature modulated DSC (TMDSC) and oscillating calorimetry.

There are a number of additional signals that can be obtained from an MT-DSC experiment that are not available from a standard DSC data set, and again there are a number of different names according to manufacturer. In this chapter the terms used are as originally described by Reading *et al.*[2]

6.3 Principles of the Technique

The temperature programme in a standard DSC experiment is given by:

$$T = T_0 + \beta t \tag{6.1}$$

where T is temperature, T_0 is the start temperature, β is the heating rate and t is time.

The response of the sample to the heating rate can be described as follows:

$$\frac{dq}{dt} = C_p \frac{dT}{dt} + f(t, T) \tag{6.2}$$

where it can be seen that the sample response is a sum of a component relating to the heat capacity of the sample and a component that is related to chemical or physical transitions. The MT-DSC experiment allows the two components to be separated as long as the experiment and calibration have been performed correctly. This allows the visualisation of the heat capacity change associated with processes such as curing and glass transitions to be determined separately from accompanying enthalpy changes.

In a modulated DSC experiment, the modulated temperature programme is given by:

$$T = T_0 + \beta t + B \sin \omega t \tag{6.3}$$

where B is the amplitude of the applied modulation in the heating rate and ω is the angular frequency of the modulation. The heat flow in an MT-DSC is given by eqn (6.4),[4] which is essentially the same as that for conventional DSC with the addition of a component relating to any lag between the applied modulation and the sample response:

$$\frac{dq}{dt} = C_p(\beta + B\omega \cos \omega t) + f(t, T) + C \sin \omega t \tag{6.4}$$

The heating rate (dT/dt in the conventional DSC heat flow equation) is denoted by $(\beta + B\omega \cos \omega t)$, where β is the underlying linear heating rate; $f(t, T)$ is the underlying kinetic function once the effect of the sine modulation has been subtracted; C is the amplitude of the kinetic response to the sine wave modulation.

At points in the MT-DSC analysis where the sample is undergoing no kinetic events then the heat flow is attributed to changes in heat capacity only. As shown in eqn (6.4), this response is due to the cosine component of the modulation. The sample response to the modulation, the modulated heat flow, will therefore be exactly ($\pi/2$ rad) or $90°$ out of phase with the modulated temperature (eqn (6.3)). In this ideal case, the modulated temperature and modulated heat flow are in-phase and the phase lag between them will be $0°$. In reality, this is not observed, see Section 6.3.2.

6.3.1 Simple Deconvolution

It is possible to deconvolute the heat flow to enable access to the C_p component of the data using a simple or a more complex method (see Section 6.3.3). It is this ability to obtain C_p data directly in one experiment that is of most interest for users of MT-DSC.

By determining the average of the sample response over one or more whole modulations, the response of the sample to the underlying linear heating rate can be determined. The data obtained in this signal will be equivalent to a linear DSC experiment at the same rate, and are commonly referred to as the "total" heat flow signal in a MT-DSC experiment. The same deconvolution can be applied to the heat input and the temperature signals to give the underlying heating rate and temperature at any time.

Comparison of the modulated component of the heat flow and the heating rate allows the heat capacity, C_p, to be determined using eqn (6.5), where A_{HF} refers to the amplitude of the modulation in the heat flow out of the sample detected by the instrument and A_{HR} refers to the amplitude of the modulation in the heating rate applied to the sample by the instrument, which is the same as $B\omega$ in eqn (6.4).

$$C_p = \frac{A_{HF}}{A_{HR}} \tag{6.5}$$

Multiplication of C_p by the underlying heating rate allows the reversing component of the heat flow to be determined. Signals that appear in the reversing heat flow signal are thermodynamically reversible at the time and temperature at which they are detected. All other events will appear in the non-reversing component, which is calculated by subtraction of the reversing component from the total heat flow (Figure 6.1).

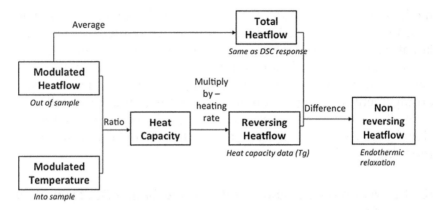

Figure 6.1 Schematic illustration showing simple deconvolution of MT-DSC data.

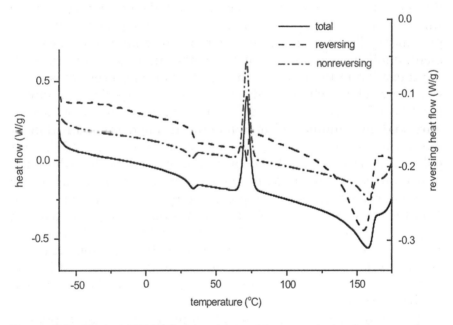

Figure 6.2 Typical MT-DSC data obtained from analysis of an amorphous sugar and protein formulation.

Examples of such thermodynamically non-reversible transitions are the endothermic relaxations that accompany glass transitions and the subsequent recrystallisations in the analysis of amorphous materials as can be seen in Figure 6.2, depicting the modulated heat flow data across the T_g, recrystallisation and subsequent melt/decomposition of an amorphous sugar sample. Upon heating through the T_g, an increase in heat capacity causes an increase in amplitude of the modulated heat flow output. Melting transitions give rise to regions with increased amplitudes in the modulated output, because of the latent heat of melting released. For low molecular weight molecules, an increase in modulation amplitude is often seen in the region between recrystallisation and melting, because of meta-stable re-arrangement causing simultaneous melting and recrystallisation. This effect can be seen as an endothermic deflection in the reversing signal because of melting, and an exothermic deflection in the non-reversing signal.

6.3.1.1 Assumptions Associated with the Simple Deconvolution Procedure

In order for this procedure to be valid, it must be assumed that the temperature excursions from the underlying heating rate are small so that the response of the process being analysed is ap-proximately linear over the small temperature interval of measure-ment. It follows that there should be sufficient modulations through the transition of interest that the deconvolution procedure is valid. In order to check whether this is the case, the experimental modu-lated heat flow is plotted against temperature in the analysis soft-ware and the number of modulations through the transition of interest is counted. Reading and Hourston[5] have suggested that for polymer systems, a useful rule of thumb is to check that where the transition is a peak in dq/dt then there should be at least five modulations over the period represented by the width at half height. They further suggested that for transitions that are represented by a step change then there should be at least five modulations over that part of the transition where the change is most rapid. It might also be useful to perform a Lissajous analysis as described in Sec-tion 6.5 to check the linearity of the response to the modulated temperature input.

This simple deconvolution is only accurate where the C term in eqn (6.4) is so small as to have no effect on the data; researchers have shown that for many commonly studied polymer systems, the discrepancy between C_p determined by data from the simple and complex deconvolution is less than 1%.[6]

6.3.2 The Phase Angle

If the phase angle is plotted with respect to temperature (Figure 6.3), two features are apparent. Firstly, that the phase angle is constant

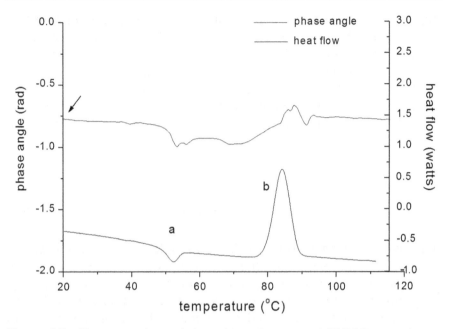

Figure 6.3 Phase angle variation throughout an MT-DSC experiment (freeze-dried sucrose).

but non-zero in the regions where no thermal events are observed (*y*-intercept indicated by the arrow) and secondly, that there are deviations from this line through the transitions labelled *a* and *b*, which are a T_g and recrystallisation respectively. The experimental effect is caused by factors such as non-ideal contact between the sample and pan and the pan and heating disk. This can be corrected for by subtraction of the constant non-zero deviation but the deviations at transitions will remain. These may be helpful in identifying glass transitions that occur in "noisy" temperature regions, such as in aqueous sugar solutions or other mixed amorphous systems. However, they also show that a sine component is contributing to the modulated heat flow as shown in eqn (6.4). This contribution is not taken into account during the simple deconvolution procedure.

6.3.3 Complex Deconvolution Using Phase Angle Correction

To remove the contribution of the sine component to the heat capacity that is used to calculate the reversing and non-reversing signals, it is necessary to calculate exactly what the contribution is from the phase angle. This method is described in more detail in the first chapter of Reading and Hourston's excellent book on the topic.[7] Briefly, the heat capacity can be described as a complex heat capacity. The phase lag is now used to divide further the complex C_p into two components: the in-phase (reversing) heat capacity (or C_p') and the out-of-phase (kinetic) heat capacity (C_{pk} or C_p'').

$$C_{pR} = C_p{}^* \cos \omega \qquad (6.6)$$

$$C_{pK} = C_p{}^* \sin \omega \qquad (6.7)$$

The total heat capacity is now referred to as the complex heat capacity, $C_p{}^*$, and the reversing and kinetic heat capacities are denoted C_p' (or reversing heat capacity, C_{pR}) and C_p'' (or kinetic heat capacity, C_{pk}), respectively, and are calculated as shown in eqn (6.6) and (6.7). The C_{pR} signal can then be used to calculate the reversing heat flow without contribution from the kinetic heat capacity, Figure 6.4.

It can be shown, however, that signals calculated from either the in-phase (complex method) or the total heat capacity (simple method) are not significantly different during the T_g or crystallisation region, the principal area analysed using this technique, although there are deviations during melting (Figure 6.5).

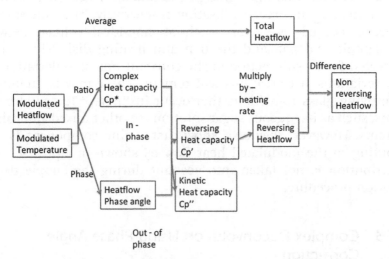

Figure 6.4 Calculation of the heat flow signals in an MT-DSC experiment including phase angle correction.

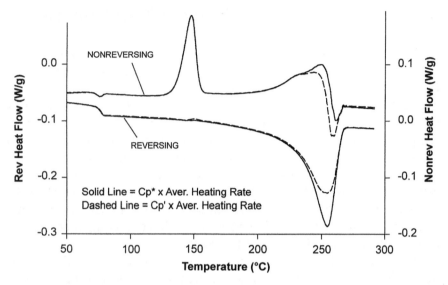

Figure 6.5   Variation of MT-DSC signals across transitions according to whether the C_p^* or C_p' is used in their calculation.[22] With kind permission from Springer Science and Business Media.

6.4 Instrumentation Design

As MT-DSC is essentially a software modification of the conventional DSC technique; in principle, any differential scanning calorimeter can be controlled in such a manner as to allow the necessary oscillation to be superimposed onto the heating rate. There is little, if any, practical difference to the data acquired from heat flux or power compensation DSCs (see Chapter 5). In recent years, improvements have been made to the cell design to reduce the mass of the furnace, while some manufacturers have also modified the instrument design to include an additional thermocouple between the sample and reference positions. This may be an advantage for situations where very accurate heat capacity information is required such as where heat capacity changes are very slow. Older (>10 years) cell designs may be restricted to low, <3 °C min^{-1} heating rates because of the relatively large mass of their furnaces. Newer calorimeters are capable of performing modulated heating profiles at rates of 10 °C min^{-1} or higher. If in doubt, it is best to refer to the manufacturer guidelines for the particular instrument that is being used.

6.5   Experimental Considerations/Best Practice

As for linear DSC experiments, it is necessary to ensure that the instrument is correctly calibrated for the heating rate, gas and pan type. If large ranges in sample mass are to be compared, it is prudent to check the effect of mass on the quality of modulation into and out of the sample, for example by plotting the modulated heat flow against temperature. Lissajous plots are the curves formed by combining two mutually perpendicular harmonic motions, and are used to investigate the response of a system to an applied modulation. Such figures can help determine the stability of the modulations during MT-DSC measurements and provide information about phase transitions. They are obtained by plotting modulated heat flow against the derivative of modulated temperature (Figure 6.6).

Ideally, a Lissajous plot will be a straight line, but in reality, it will be an ellipse because of the effects of heat dissipation and phase lag. If the sample response to the modulation is ideal then the ellipses will retrace. The slope of the ellipse gives information about the heat capacity; the width gives information about heat dissipation. For

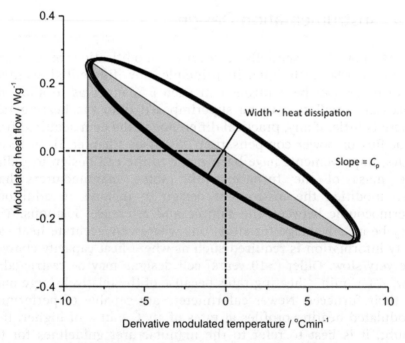

Figure 6.6   Representation of a Lissajous plot.

example, a difference will be seen between the width of the ellipses above and below the T_g because of the increased entropy in the liquid phase leading to increased heat dissipation.[8–10] Lissajous figures may also be used in order to determine if a loss of steady state parameters has occurred. An example of this would be where the modulation period is too short, causing a severe distortion of the ellipse. This method has been reported as a useful means of assessing the suitability of a modulation profile.[8]

The additional modulation parameters of temperature amplitude and period allow greater variability in the experimental design than standard DSC analyses so that it is possible to fine-tune the method to suit best the transitions being analysed and obtain more information from the analysis of one sample. However, this requires that the user has an understanding of the effect of parameters on the quality of data, in order to prevent the introduction of experimental artefacts. For example, there are practical problems associated with the measurement of events that occur in the region of state changes, especially the sharp melting transitions associated with small organic molecules or metals because of the evolution or requirement of latent heat to form or break the bonds involved. This gives rise to problems with the applied modulation and also means that it can no longer be assumed that the response of the process being analysed is linear or that the temperature excursions from the underlying heating rate are small. Therefore, the deconvoluted signals will contain artefacts (Figure 6.7).

It has often been said that one of the great advantages of MT-DSC is that it can be switched off. For a great many samples, there may be no benefit to running a MT-DSC experiment over a conventional DSC scan. DSC scans are faster and easier to interpret if all that is required is confirmation of a melting transition.

Even if you are sure that MT-DSC is going to be useful for your sample, it is always beneficial to perform a DSC scan to determine whether there is any benefit to using MT-DSC. It is also advisable to run a sample over the same range and at the same heating rate in a TGA to check for any undesirable decomposition. TGA furnaces are better designed to cope with the decomposition products than DSC cells. As MT-DSC experiments are generally slower than DSC, it may be desirable to choose a smaller temperature range for a MT-DSC experiment that centres on the transition of interest. That said, it is useful to ensure that there is sufficient baseline on either side of any transition to ensure that the analysis software is able to determine the difference between baseline and transition. It is worthwhile

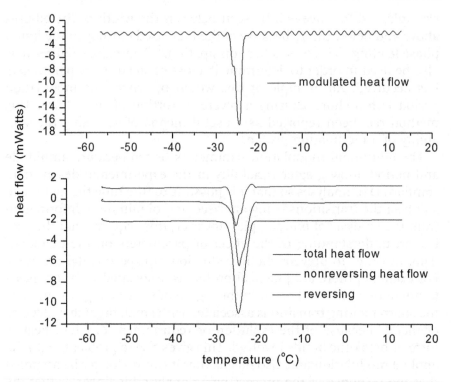

Figure 6.7 MT-DSC trace of the melt of *n*-decane (signals have been offset for clarity).

checking to see that the modulation has "settled" below the region of interest, although for newer calorimeters, this will be less of an issue.

For many amorphous samples, the role of water or other plasti-cizers can be extremely important. It is useful to think about what will be done with the data generated. If the T_g is to be used to predict the storage stability of a product then the role of the sample pan in en-abling volatiles to be lost or trapped should be understood. Similarly, heat–cool–heat runs, whereby samples are heated through a T_g then cooled before reheating, can be useful to confirm a T_g. If pin-holed pans are used then this may lead to the removal of peaks associated with the loss of water from the sample that will also have the effect of making it easier to observe the T_g on the reheat. However, this may also have the effect of increasing the T_g temperature since water will be lost from the sample on the initial heat.

Formation of amorphous material by quench cooling from the melt can be a useful way of producing amorphous material in the DSC pan, especially if it is desirable to increase thermal contact between sample and pan, but it is then also prudent to check for any decomposition that may have occurred during melting, for example by analysing a sample heated in the same manner in a TGA.

6.5.1 Improving the Accuracy of Experimental Data

The considerations discussed in Chapter 5 regarding sample preparation will also be valid for MT-DSC experiments. Small sample sizes of uniform size distribution sealed into hermetic pans minimise the effects of packing, thermal gradients and heat transfer due to convection into the air. Hermetic pans can distort during analysis if the sample loses volatiles or decomposes during analysis. This will affect the thermal conductivity between the sample and the heating disc, giving rise to artefacts in the baseline and apparent changes in heat capacity. Small samples also increase resolution, but lower sensitivity. Large samples can be helpful in broadening small transition peaks but this effect can give rise to ghost peaks because of thermal gradients in the sample. Low heating rates are usually employed to minimise the effects of lag and, more importantly, loss of steady state, since this is presumed in the basic MT-DSC equations. This increases resolution, but lowers sensitivity. It is advisable to match the reference and sample pan masses so that differences in pan mass are avoided. In cases where baseline noise must be kept to a minimum, an inert mass may be added to the reference pan (for example, pieces of pan-lid) to match the mass of the sample.

There will generally be a "running in zone" at the start of a MT-DSC experiment, during which the modulation is not exactly as programmed. Again, inspection of the modulated heat flow against temperature will reveal this region. For this reason, it is useful to ensure that the starting temperature is well below the temperature region of interest for the sample undergoing analysis. For analysis of data caused by low energy transitions or those that are accompanied by very small changes in C_p, it can be useful to really zoom in on the data before selecting the baseline to ensure that there is both enough baseline on either side and also that the baseline is of a good quality, *i.e.* that it does not also include the start or end of another transition.

6.6 Calibration

In addition to the calibration required for a standard DSC experiment (Chapter 5), it is usual to perform a heat capacity calibration procedure. This entails running a sample of a standard material such as aluminium oxide (sapphire).

Some instruments allow users to save a calibration file across the temperature range of subsequent experiments. Others allow users to enter a single point value, which is generally selected to be in the centre of the temperature region of interest. The user calculates the heat capacity constant by comparing the experimentally determined C_p with a literature value at the same temperature. For many users who simply wish to identify that an amorphous phase is present, this may well be sufficient.

If an instrument only allows input of a single point calibration but there is the need to work more accurately, then it may be necessary to export the uncalibrated data file into a spreadsheet in order to manipulate the data to obtain the calibrated data. This is possible because an equation has been developed that can generate a set of theoretical values for the sapphire C_p across the temperature range of interest using eqn (6.8).[9]

$$C_P = 8.6446 - (0.4929T) + (8.2336 \times 10^{-3}T^2) - (3.5739 \times 10^{-5}T^3)$$
$$+ (7.556 \times 10^{-8}T^4) - (8.0336 \times 10^{-11}T^5) + (3.4219 \times 10^{-14}T^6) \qquad (6.8)$$

Comparison of the empirical with the literature values gives the required constant. This is important even if heat capacity is not being determined in the MT-DSC experiment since the reversing signal is calculated using the heat capacity, which in turn is used to determine the non-reversing component. Full details of this procedure are given in Reading *et al.* (2007).[10]

It has been found that long period times and large modulation amplitudes give the most accurate heat capacity measurements.[8] One scan only is sufficient to obtain heat capacity values using MT-DSC (two would be required for the equivalent conventional DSC experiment) if a heat capacity calibration has been performed using the same conditions. As for the standard DSC experiments, calibration should be checked regularly and after any contamination/cleaning as well as when heating rate; modulation; gas; pan and/or sample type are changed.

6.7 Applications

There are a number of reviews and books that cover the use of MT-DSC for various applications. For details of some of the more commonly used applications, the reader is referred to the further reading section below. By way of example, the most commonly used applications are given next.

6.7.1 Glass Transitions, T_g

MT-DSC is advantageous in the measurement of glass transitions because the deconvolution procedure separates the step change in the heat capacity that appears in the reversing signal from the accompanying relaxation in the non-reversing signal (Figure 6.8).

This allows for more accurate determination of T_g values and of the magnitude of the enthalpic relaxation itself. Theoretically, this removes the need for lengthy annealing or cycling. It also makes it possible to resolve overlapping events such as glass transitions and recrystallisations in a mixture. Weak glass transitions can be detected more easily because the sine wave increases the sensitivity and, to an extent, the curvature of baseline effects is removed, so that samples containing diluents or having low amorphous contents can still be accurately analysed. However, the large modulation amplitudes and low scanning rates that must be employed to give

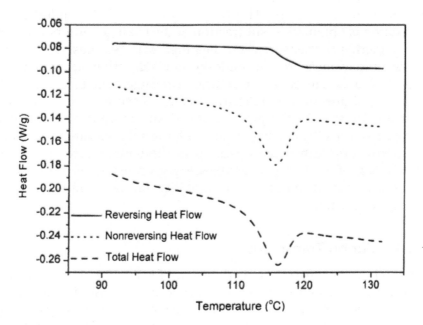

Figure 6.8   The T_g of lactose observed with MTDSC.[12]
With kind permission from Springer Science and Business Media.

both high resolution and sensitivity can cause problems because of the requirement for a sufficient number of modulations throughout the course of one thermal event as described in Section 6.5.1. This requires low heating rates that increase the length of experiments.

Care must be taken if very accurate T_g measurements are required because the temperature at which the transition occurs is dependent on the heating rate, beta and modulation frequency, ω, with lower frequency rates decreasing the onset of the T_g upon heating.[11] Hysteresis effects can occur if the cooling and heating rates are not matched. This effect is important when determining the enthalpy of an endothermic relaxation accompanying a T_g. Often a T_g determined from the reversing heat flow will be shifted slightly compared with the same transition determined from the total heat flow. This is because the transition measured is kinetic and is therefore affected by the applied frequency. This effect also gives rise to a peak in the non-reversing signal.

To calculate the effect of the peak in the non-reversing signal caused by the frequency effect, it is necessary to measure the enthalpy twice, once upon heating (which will also contain the frequency effect) and once upon recooling using the same modulation parameters. The true magnitude of the endothermic relaxation is obtained by subtraction of the cooling endotherm from the endotherm measured on heating. MT-DSC has been used to accurately measure the endothermic relaxation accompanying the T_g that is caused by structural rearrangement of the glass.[12] The magnitude of the relaxation is a measure of the mobility that was present in the glassy state. If an amorphous material is stored below its T_g, the size of the endothermic relaxation obtained upon reheating can be correlated with the temperature of storage. If a series of such experiments are performed, it is possible to determine the dependence of mobility on temperature in the amorphous state.

6.7.2 Curing Transitions

Van Mele's group first described the use of MT-DSC to investigate the isothermal cure and vitrification process in thermosetting systems[13] and went on to describe the use of the technique for probing reacting polymer systems[14] and phase separation in miscible polymer blends.[15] This group continues to publish widely on the application of MT-DSC and other thermal techniques to polymer systems.

6.7.3 Crystallisation and Melting Phenomena Associated with Small Organics

The standard DSC method is well suited to the determination and analysis of melting transitions, while fast scanning DSC may be particularly useful for investigation of the state transitions and interconversion processes associated with metastable systems. However, great care must be taken when attempting to use MT-DSC for analysis of melting transitions as detailed in the section on phase angle (Section 6.3.2). For small organic molecules, the melt is generally a transition with a narrow temperature range. This can give rise to significant deviations from linearity of sample response to the applied modulation, which invalidate the assumption of linear response implicit in the initial derivation of the total heat flow signal. Polymers generally exhibit wider melting transitions because of the range of molecular weights and MT-DSC can be useful to obtain information about the degree of crystallinity. It is therefore useful to remember the definition described by Reading *et al.*[7] that "the term non-reversing was coined to denote that at the time and temperature the measurement was made the process was not reversing although it might be reversible." So processes that may well be reversible given sufficient time or change in temperature, such as crystallisation or loss or volatiles, are indeed non-reversible under the conditions of the MT-DSC experiment.

6.7.4 Polymer Systems

The group of Wünderlich produced much excellent research investigating how MT-DSC can be used to probe reversible and irreversible changes in macromolecular systems. For semi-crystalline systems, they have been described as having multiple glass transitions attributable to rigid amorphous fractions and glass transitions within crystals, and also locally reversible melting at the surface of chain-folded crystals.[16] The Wünderlich group was responsible for setting up the Advanced THermal Analysis System (ATHAS) databank, which is a useful resource for polymer scientists.

6.7.5 Quasi-Isothermal MT-DSC

Quasi-isothermal (QI) MT-DSC experiments can be used to investigate the effect of temperature on heat capacity, by recording sets of isothermal data through the region of interest. By using very small

increments in temperature, the effect of heating rate can be substantially removed. In a QI MT-DSC experiment, the temperature is
modulated about a set underlying temperature for a period of time
before increasing the temperature incrementally to another temperature and repeating the collection, which results in a set of quasi-
isothermal data. The isothermal periods should be long relative to the
increment between steps to reduce the effect of heating rate.[17] This
design has been used to investigate polymorphic transformations,
such as the enantiotropic conversion between caffeine form II and
form I upon heating.[18,19] Figure 6.9 shows an example of a QI-MT-
DSC data set, in this case the heat capacity output obtained by heating
through the enantiotropic conversion between caffeine form II and
form I.

QI-MTDSC experiments can also be used for accurate determination of T_g values. For this application, the experiment is started at a
temperature above the T_g region and the increments are selected to
cool through the T_g. Before collecting data during the quasi-isothermal experiment, it is necessary to remove annealing effects by
first cooling and reheating. The temperature is modulated without

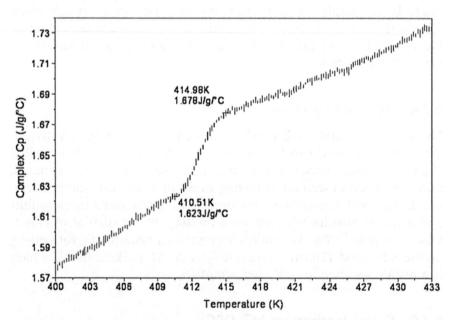

Figure 6.9   Quasi-isothermal MT-DSC heat capacity data for caffeine
form II with modulation ±0.75 K over 80 s; increment 0.2 K;
isothermal period 30 min (one of five replicates shown).[21]
Reproduced from ref. 18 © 2007 Wiley-Liss, Inc. and the
American Pharmacists Association.

ramping; the first ten minutes at each temperature are used to allow the sample to equilibrate and so are not saved.[20,21] The QI-MTDSC technique has also been used to investigate the isotropization transition between the liquid crystalline and liquid state.[17]

6.8 Summary

MT-DSC is a useful software modification to the standard DSC used in many instrument laboratories. Although it is most often associated with identification of hard to detect glass transitions, it has application to a wider range of material properties.

Knowledge of the assumptions associated with the response of the sample to the applied modulation allows the user to make informed choices about appropriate experimental design and opens up the possibility of accessing a wider range of data that can be used to investigate thermodynamic and kinetic behaviour of materials.

6.9 Further Reading

The following book was cited in the text but is worth highlighting separately:

Modulated-Temperature Differential Scanning Calorimetry: Theoretical and Practical Applications in Polymer Characterisation, ed. M. Reading and D. J. Hourston, Springer, 2006, vol. 6.

The ASTM web-site has links to the various ASTM methods that use MT-DSC (www.astm.org) including the method for determining specific heat capacity using MT-DSC ASTM E2716-09(2014).

References

1. M. Reading, D. Elliott and V. Hill, in *Proceedings of the 21st NATAS Conference*, North American Thermal Analysis Society, Atlanta, GA, 1992.
2. M. Reading, *Trends in Polym. Sci.*, 1993, **1**(8), 248.
3. T. Lever, P. Haines, J. Rouquerol, E. L. Charsley, P. Van Eckeren and D. J. Burlett, *Pure Appl. Chem.*, 2014, **86**, 545.
4. M. Reading, D. Elliott and V. Hill, *J. Therm. Anal.*, 1993, **40**, 959.
5. *Modulated Temperature Differential Scanning Calorimetry*, ed. M. Reading and D. J. Hourston, Springer, 2006.
6. M. Reading, *Thermochim. Acta*, 1997, **292**, 179.

7. D. M. Price, A. A. Lacey, M. Reading, Theory and Practice of Modulated Temperature Differential Scanning Calorimetry, in *Modulated Temperature Differential Scanning Calorimetry Theoretical and Practical Applications in Polymer Characterisation*, ed. D. J. Hourston and M. Reading, Springer, New York, 2006, vol. 6, pp. 1–81.

8. V. L. Hill, D. Q. M. Craig and L. C. Feely, *Int. J. Pharm.*, 1999, **192**, 21.

9. D. G. Archer, *J. Phys. Chem.*, 1993, **22**, 1441.

10. D. C. M. Reading, J. R. Murphy, V. L. Kett, in *Thermal Analysis of Pharmaceuticals*, ed. D. C. a. M. Reading, CRC Press, 2007, p. 102.

11. A. Boller, C. Schick and B. Wunderlich, *Thermochim. Acta*, 1995, **266**, 97.

12. M. Barsnes, D. Q. M. Craig, P. G. Royall and V. L. Kett, *Pharm. Res.*, 2000, **17**, 696.

13. G. Van Assche, A. Van Hemelrijck, H. Rahier and B. Van Mele, *Thermochim. Acta*, 1997, **268**, 121.

14. S. Swier, H. Van Hemelrijck, E. Verdonck and B. Van Mele, *J. Therm. Anal. Calorim.*, 1998, **54**, 585.

15. G. Dreezen, G. Groeninckx, S. Swier and B. Van Mele, *Polymer*, 2001, **42**, 1449.

16. B. Wunderlich, *Pure Appl. Chem.*, 2009, **81**, 1931.

17. W. Chen, M. Dadmun, G. Zhang, A. Boller and B. Wunderlich, *Thermochim. Acta*, 1998, **324**, 87.

18. R. Manduva, V. L. Kett, S. R. Banks, J. Wood, M. Reading and D. Q. M. Craig, *J. Pharm. Sci.*, 2008, **97**, 1285.

19. S. Qi and D. Q. M. Craig, *Mol. Pharm.*, 2012, **9**, 1087.

20. P. G. Royall, D. Q. M. Craig and C. Doherty, *Pharm. Res.*, 1998, **15**, 1117.

21. L. C. Thomas, A. Boller, I. Okazaki and B. Wundrlich, *Thermochim. Acta*, 1997, **291**, 85.

22. S. R. Aubuchon and P. S. Gill, *J. Therm. Anal.*, 1997, **49**, 1039.

7 Isothermal Microcalorimetry

Simon Gaisford

UCL School of Pharmacy, University College London, 29-39 Brunswick
Square, London, WC1N 1AX, UK
Email: s.gaisford@ucl.ac.uk

7.1 Introduction and Principles

An inevitable consequence of the laws of thermodynamics is that if a
material can change from its current state to a more stable state then
it will do so. This may involve a change in physical or chemical form
or chemical reaction with another compound. Diamond, for example,
is the metastable form of carbon and over time, it will convert to
graphite. The only consideration to be taken into account in determining if such change is of importance is length of time over which it
will occur. In the case of the transition of diamond to graphite, the
process will take millions of years and so it may be ignored. For other
materials, the rate of change may be more pronounced. For manufactured materials, it is generally the case that the initial state is the
one with the desired properties and over time, changes in the material
will lead to deterioration in its properties and hence, by definition, a
loss of quality. The measurement and quantification of change is
therefore critically important when setting product lifetimes.

Whenever an analytical measurement is made in order to determine whether there has been a change in a sample, two conditions
must be satisfied. Firstly, there must be a change in some physicochemical property of the sample and secondly, the analytical

Principles of Thermal Analysis and Calorimetry: 2nd Edition
Edited by Simon Gaisford, Vicky Kett and Peter Haines
© The Royal Society of Chemistry 2016
Published by the Royal Society of Chemistry, www.rsc.org

instrument must be sensitive to that change. Calorimeters are unique in the sense that the property they measure, heat (or enthalpy), is ubiquitous and so is a *universal accompaniment to chemical or physical change*. In broad terms, this means that calorimeters are, in principle, capable of detecting change in any sample. The only limitations are the detection limit of the instrument and the size of the sample[†]. This versatility has led to the use of calorimetry in a multitude of areas, from characterisation of simple molecules to monitoring the processes of life itself. It is also often the case that calorimeters are sensitive to subtle changes in a material that are not apparent with other analytical techniques.

Calorimeters can be divided into those that heat the sample during measurement (differential scanning calorimetry, DSC) and those that operate isothermally (isothermal microcalorimetry, IMC)[‡]. It follows that DSC is sensitive to *thermally-driven* phase transitions while IMC is sensitive to *time-dependent* phase transitions or processes. The two techniques are therefore complementary and together form a powerful suite that permits characterisation and investigation of a multitude of materials. The principles and main applications of DSC are discussed in Chapter 5. Here, the focus is on the principles, experimental arrangement and application of isothermal microcalorimetry. One point to note here is that isothermal microcalorimeters measure heats on the microwatt scale or below. The isothermal reaction calorimeters discussed in Chapter 8 record powers on the milliWatt scale and above. The principles of measurement are similar, but their areas of application are different.

7.2   Definitions and Nomenclature

The word calorimetry derives from calor (Greek, heat) and metry (Latin, measurement) and so literally means the measurement of heat. It is one of the oldest analytical techniques, tracing its roots back to the ice calorimeters of Black and Lavoisier and Laplace (*ca.* 1780).[1] In these early designs, a sample is surrounded by ice. As the sample reacts, and so long as the process is exothermic, heat is exchanged with the ice, resulting in melting and formation of water. Knowledge of the latent enthalpy of fusion of water permits the total heat released during measurement to be calculated simply by

---

[†]Although it is possible to construct a calorimeter to enclose virtually any sample.
[‡]Change in pressure may also be used, but pressure-perturbation calorimetry is not considered here.

measuring the volume of water produced. If, in addition, the rate of production of water droplets is observed, then information on the kinetics of the process can be inferred. It follows that IC data contain information on both the thermodynamics (heat, q, given the SI unit of Joules, J[§]) and the kinetics (power, dq/dt, given the SI unit of Watts, W) of the system under investigation.

In one of their first experiments, Lavoisier and Laplace[1] measured the heat of combustion of carbon, finding that "one ounce of carbon in burning melts six pounds and two ounces of ice". This allowed a value for the heat of combustion of -413.6 kJ mol^{-1} to be calculated. When compared with the most accurate current value of -393.5 kJ mol^{-1}, it can be seen that the ice calorimeter was remarkably accurate. In a later experiment, a guinea pig was placed in the sample cell. By comparing the heat evolved from the guinea pig with the amount of oxygen consumed, they concluded:

> ...*respiration is thus a combustion, to be sure very slow, but otherwise perfectly similar to that of carbon; it takes place in the interior of the lungs, without emitting visible light, because the matter of the fire on becoming free is soon adsorbed by the humidity of these organs. The heat developed in the combustion is transferred to the blood which traverses the lungs, and from this is spread throughout all the animal system*

This conclusion is remarkably astute, given the limited number of data on which it is based, and shows the power of the calorimetric technique.

Modern instruments have not actually changed significantly since these early designs, save for the fact that modern electronics allow for the construction of apparatus that can measure both exothermic and endothermic events. In addition, most instruments operate with an inert reference material, which is used to correct for background noise and environmental fluctuations (and so are *differential*). An example of a typical commercial system is shown in Figure 7.1.

A large number of isothermal calorimeters are commercially available. They are based on a number of different operating principles and use a variety of nomenclatures. Unfortunately, there is no common agreement on the naming of instruments and this often leads to confusion.[2] Even the term isothermal microcalorimetry is confusing, as it is unclear whether the instrument is measuring heat

---

[§]The late Tom Hofelich was once asked why he presented all his data in calories, rather than Joules. He replied "When calorimeters are termed Joulerimeters, I will present my data in Joules."

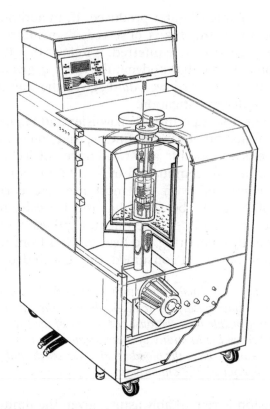

Figure 7.1 Schematic view of an isothermal microcalorimeter (image courtesy of TA Instruments LLC).

on a micro-Watt scale (the meaning in this chapter) or if the sample size required for measurement is on the order of micrograms. Instruments may also be named after their designer or an operating characteristic (Black's Ice Calorimeter, Parr's Oxygen Bomb Calorimeter, Gas Perfusion Calorimetry and so on), which again reveals no details of its operating principles.

An excellent example of the spread of names used by authors to describe their calorimeters is provided by Hansen,[2] who noted the following names in a special issue of *Thermochimica Acta* on developments in calorimetry: levitation melting calorimetry, IC-calorimetry, flow-calorimetry, water absorbed dose calorimetry, microcalorimetry, DSC analysis, temperature-modulated DSC, high temperature calorimetry and nanocalorimetry. He suggests that, in the way common names for chemicals have been replaced with IUPAC sanctioned nomenclature, names for calorimeters and calorimetric procedures should be replaced or supplemented with a systematic nomenclature that gives a clear indication of the method and mode of operation of the calorimeter used, although this is some way away and common or trade names will probably never disappear.

7.3 Principles of the Technique

Irrespective of the problems caused by the lack of a common no-menclature, it is important to know the basic designs and operating principles that underpin modern calorimeters in order to understand the origin of calorimetric signals, to be able to construct the most appropriate experimental arrangement and to draw comparison between them. There are only three ways by which heat may be measured experimentally, so all calorimeters are based on one of the following principles:

7.3.1 Power-compensation Calorimetry

The power required to maintain isothermal conditions in a calorimeter is recorded directly, the power being supplied by an electronic temperature controller in direct contact with the calorimeter. In practice, power compensation calorimeters are usually designed to accommodate both a sample and an inert reference and the difference in power (ΔP) supplied by the element to the two sides is measured. The power output from the sample is the inverse of ΔP supplied by the element. In order to be able both to heat and cool, the element is usually of the Peltier type. Power-compensation designs are often used for DSC instruments, as they permit rapid changes in temperature, but they also feature in some isothermal instruments (in particular, those designed for isothermal titration calorimetry).

7.3.2 Adiabatic Calorimetry

In an ideal adiabatic calorimeter, there is no heat exchange between the calorimetric vessel and its surroundings (this is usually attained by placing an adiabatic shield around the vessel). Thus any change in the heat content of a sample (q) as it reacts causes either a temperature rise (exothermic processes) or fall (endothermic processes) in the vessel. The change in heat must be equal to the product of the measured temperature change (ΔT_{exp}) and an experimentally determined proportionality constant (or calibration constant, ε):

$$\Delta T_{exp} = \frac{q}{\varepsilon} \tag{7.1}$$

The value of ε is determined by electrical calibration (see Section 7.6.1). Ideally, the value of ε should equal the heat capacity of the calorimeter vessel (C_v, the vessel including the calorimetric ampoule, block, heaters, thermopiles and the sample), but in practice, losses in heat because true adiabatic conditions cannot be attained mean the value will differ slightly. However, assuming the losses are the same for sample and reference, the heat returned will be accurate and is calculated as:

$$q = \varepsilon \Delta T_{exp} \qquad (7.2)$$

It is apparent from the definition earlier that C_v depends on the sample, which affects the sensitivity of the instrument. A smaller heat capacity results in a larger rise in temperature for a given quantity of heat and, consequently, better sensitivity. However, calorimeters with low heat capacities are more sensitive to environmental temperature fluctuations and have lower baseline stabilities. Any calorimeter design therefore results in a compromise between baseline stability and measurement sensitivity.

Upon completion of a reaction in an adiabatic calorimeter, the change in temperature will have been recorded. It is assumed that the temperature change arises solely from the event occurring in the vessel (ΔT_{ideal}) but in practice, other events, such as ampoule breaking and mechanical stirring, contribute to the temperature change of the vessel. Further, since true adiabatic conditions are nearly impossible to maintain, there will be heat-leak from the calorimeter. When the heat-leak is factored into the system, the calorimeter is said to operate under semi-adiabatic (or isoperibolic) conditions. Corrections are then necessary in order to return accurate data. The basic assumption is that:

$$\Delta T_{exp} = \Delta T_{ideal} + \Delta T_{other} \qquad (7.3)$$

where ΔT_{other} is the contribution to the temperature change from processes not related to the sample. Usually the method of Regnault–Pfaundler, which is based on the dynamics of the break, is used to determine the value of ΔT_{adj}:[3]

$$\Delta T_{other} = \int_{t_{start}}^{t_{end}} \frac{1}{\tau} (T_\infty - T) dt \qquad (7.4)$$

where T is the temperature of the vessel and its contents at time t, T_∞ is the temperature that the vessel would attain after an infinitely long time period, t_{start} and t_{end} are the start and end times of the experiment respectively and τ is the time constant of the instrument. Note that T_∞ is effectively the value of T at t_{end} and is commonly described as the steady-state temperature of the vessel. The values of T_∞ and τ are calculated by analysis of the baseline sections immediately before and after the reaction. These baseline sections will be approaching the temperature of the surroundings exponentially and are described by:

$$T = T_\infty + (T_0 - T_\infty)e^{\frac{-t}{\tau}} \qquad (7.5)$$

The experimental data are fitted to eqn (7.5) using a least-squares minimising routine to return values for T_∞ and τ. Once these are known, ΔT_{other} can be calculated from eqn (7.4) and ΔT_{ideal} from eqn (7.3).

7.3.3 Heat Conduction Calorimetry

The temperature difference across a path of fixed thermal conductivity is measured. In a typical arrangement, sample and reference vessels are maintained at a constant temperature in the calorimeter; any heat released or absorbed by a reaction is quantitatively exchanged with a surrounding heat-sink *via* an array of thermocouples (a thermopile). The thermopile generates an electrical potential (U) that is proportional to the heat flowing across it. This is multiplied by an experimentally determined calibration constant to give power:

$$\Phi = \frac{dq}{dt} = \varepsilon U \qquad (7.6)$$

The total change in heat is given by integrating the power data with respect to time:

$$q = \varepsilon \int U dt \qquad (7.7)$$

The sensitivity of the calorimeter will depend on the heat capacity of the vessel and also on the electrical potential generated by the thermopile in response to a temperature gradient (g, the Seebeck coefficient, with units of VK^{-1}):

$$U = g\Delta T = g(T - T_0) \qquad (7.8)$$

where T_0 is the temperature of the heat-sink and T is the temperature of the sample following reaction. Rearrangement of eqn (7.8) gives:

$$T = \frac{U}{g} + T_0 \qquad (7.9)$$

The time derivative of eqn (7.9) is:

$$\frac{dT}{dt} = \frac{1}{g}\frac{dU}{dt} \tag{7.10}$$

Ideally, there would be no time delay between heat change in the sample and voltage being measured by the instrument, in which case the measured electrical potential multiplied by the proportionality constant gives power directly, as stated earlier. Experimentally, however, there is usually a delay in detection (known as the dynamic response of the instrument). The dynamic response can be formalised by considering the events that must occur between heat change in a sample and that heat causing an electrical potential to be generated by the thermopiles.

Any event in the sample must cause a temperature change in the calorimeter (as in the case of adiabatic calorimetry), the magnitude of the temperature change being dependent upon the heat capacity of the vessel and its contents. A temperature gradient forms between the vessel and the heat-sink and so heat must be exchanged with the heat-sink to restore thermal equilibrium. The rate of heat transfer is dependent on the heat transfer coefficient (k in W K^{-1}) of the thermopiles. The experimentally measured power (Φ_{exp}) therefore derives from two events (temperature change and heat transfer) and can be represented by a *heat balance* equation:

$$\Phi_{exp} = \Phi_{trans} + C_v\frac{dT}{dt} \tag{7.11}$$

where the term Φ_{trans} represents heat transfer and the term $C_v\frac{dT}{dt}$ represents temperature change. Newton's law of cooling states that the rate of heat transfer is proportional to $T - T_0$:

$$\Phi_{trans} = k(T - T_0) \tag{7.12}$$

where k is the heat transfer coefficient. It follows that:

$$\Phi_{exp} = k(T - T_0) + C_v\frac{dT}{dt} \tag{7.13}$$

Equation (7.13) describes a calorimeter with a single vessel. In practice, a twin calorimeter (sample and reference) arrangement is often used. In this case, an equation of the form of eqn (7.13) can be written for both sample and reference calorimeters. Assuming the calorimeters are carefully constructed so that they have equal values of k and C_v, then it is possible to write:

$$\Phi_{exp} = k(T_s - T_r) + C_v \frac{dT_s}{dt} \tag{7.14}$$

Further consideration permits the heat balance equations to be related to the electrical potential produced by the thermopile. Rearrangement of eqn (7.9) yields:

$$(T - T_0) = \frac{U}{g} \tag{7.15}$$

It follows that:

$$\Phi_{exp} = \frac{k}{g} U + \frac{C_v}{g} \frac{dU}{dt} \tag{7.16}$$

which, upon rearrangement, gives:

$$\Phi_{exp} = \frac{k}{g} \left(U + \frac{C_v}{k} \frac{dU}{dt} \right) \tag{7.17}$$

If the following definitions are made:

$$\varepsilon = \frac{k}{g} \tag{7.18}$$

$$\tau = \frac{C_v}{k} \tag{7.19}$$

where ε is the calibration constant defined earlier and τ is the time constant of the instrument then the following can be written:

$$\Phi_{exp} = \varepsilon \left(U + \tau \frac{dU}{dt} \right) \tag{7.20}$$

Equation (7.20) is known as the Tian equation, and is the most well-known mathematical description of the relationship between experimentally measured power and electrical potential generated across a thermopile in a heat conduction calorimeter. As written, the Tian equation applies to a single calorimeter, but the same derivation applies to twin instruments.

7.3.4 Single *Versus* Multiple Calorimeters

A single calorimeter is constructed to hold only one ampoule. Because this means a measurement can be made on only one sample at a time, it is necessary to perform a blank (or reference) experiment as a control, in order to subtract any heat not generated from the process of interest. In addition, inspection of eqn (7.13) shows that single calorimeters are also affected by temperature fluctuations in the heat-sink. This means that single calorimeters are really only suited to measurements made over a short time period (so significant fluctuations in temperature do not occur). Adiabatic calorimeters are usually single vessels.

Multiple calorimeters can accommodate more than one vessel. Typically, a twin arrangement is used, with one of the vessels being used as a reference. The two calorimeters, constructed to be as closely matched as possible, are usually connected in opposition (*i.e.* an exothermic process will produce a positive signal in one side and a negative signal in the other). Equation (7.13) shows that such a design automatically corrects for temperature fluctuations and, if a suitable reference sample is available, also allows subtraction of the blank signal during an experimental measurement. Calorimeters designed for a higher throughput of samples may have capacity for more than two vessels; in this case, one will be used as a (common) reference and the others for samples.

7.4   Instrumentation Design

In addition to the basic operating principle of the instrument, the arrangement of the sample in the calorimeter is also extremely important because it defines the type of measurement that can be performed.

7.4.1   Ampoule Calorimetry

The sample is contained within a sealed ampoule placed in the measuring position of the calorimeter. The ampoule is typically constructed from metal (stainless steel or an inert alloy, such as Hastelloy) or glass and will have a lid with a hermetic seal (Figure 7.2). Maintaining an air-tight enclosure is important because evaporation of water or solvent will cause an endothermic power. Ampoules may be reusable or disposable. It is also possible to control the humidity of the atmosphere within the ampoule with the use of a mini-hygrostat containing a saturated salt solution. Ampoule calorimetry has the most general application, because virtually all types of sample can be accommodated, including solids, liquids or heterogeneous materials.

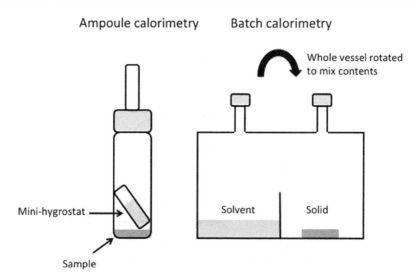

Figure 7.2   Schematic representations of ampoule calorimetry (l) and batch calorimetry (r).

7.4.2   Batch Calorimetry

The sample ampoule in the calorimeter is divided with a low partition, which allows liquid and/or solid samples to be loaded separately. Once data capture has started, the chamber is rotated, allowing the materials to mix (Figure 7.2). This arrangement is particularly suited to measuring heats of interaction and/or mixing.

7.4.3   Solution, or Ampoule-breaking, Calorimetry

Solution calorimetry is not, as the name suggests, the calorimetric study of solutions. Rather, it is similar in principle to batch calorimetry, where two materials are held in separate chambers prior to measurement. In this case, one sample (usually a solid) is encapsulated in a small, sacrificial ampoule. The other sample (usually a liquid) is held in the main sample chamber (Figure 7.3). To commence measurement, the ampoule containing the solid is broken, allowing the two materials to mix. Ampoules can be glass (often referred to as crushing ampoules) or metal (in which case the ampoule is constructed of several pieces which fall apart).

7.4.4   Flow Calorimetry

A flow system is arranged in a similar manner to the ampoule system described, but the ampoule is modified so that it is possible to

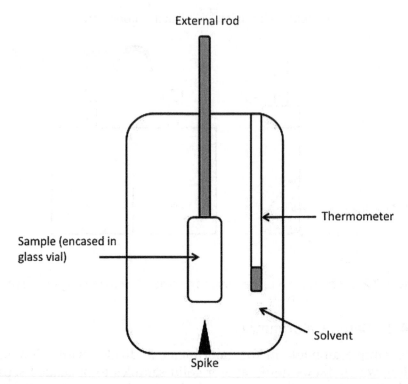

External rod

Thermometer

Sample (encased in
glass vial)

Solvent

Spike

Figure 7.3 Schematic representation of an ampoule-breaking (solution)
calorimeter.

circulate (with a peristaltic pump) a liquid through it. The system can
be arranged so that the liquid is returned to an external reservoir or
pumped to waste. It is also possible to pump two liquids and arrange
for them to mix in the sample chamber, in which case, the system is
referred to as a *flow-mix* calorimeter. One aspect of flow calorimetry
that is particularly important is to ensure the temperature of the
flowing liquid is exactly the same as that of the calorimeter. This can
be achieved with slow flow rates (2–4 mL h^{-1}) and by circulating liquid through a heat-exchange coil. The coil is usually sited in the
thermostatted bath of the calorimeter.

7.4.5 Gas Perfusion Calorimetry

Gas perfusion calorimeters are analogous to flow calorimeters, except
a gas is perfused through the calorimetric vessel instead of a liquid.
The relative humidity (RH) or relative vapour pressure (RVP) of the gas
stream is controlled during the course of an experiment. In a typical

Titration calorimetry   Gas perfusion calorimetry

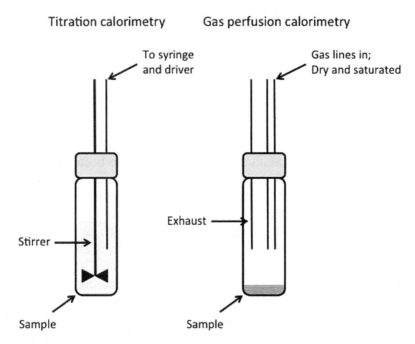

Figure 7.4   Schematic representations of titration calorimetry (l) and gas perfusion calorimetry (r).

arrangement, two gas lines are routed into the calorimeter (Figure 7.4). One is dry and the other passes through a series of reservoirs containing the vapour (usually water or ethanol, but any solvent that does not damage the instrument can be used). Proportional mixing of the flow rates of the two gas lines permits accurate control of the RH or RVP in the vessel.

7.4.6   Isothermal Titration Calorimetry

Isothermal titration calorimetry (ITC) is probably the most widely used form of isothermal microcalorimetry after ampoule calorimetry. In a typical arrangement, a syringe, mounted externally to the instrument, is loaded with solution of a ligand while the calorimeter vessel is charged with a solution of substrate (Figure 7.4). A cannula connects the syringe to the calorimeter vessel. A motor is used to drive the syringe plunger and so to inject aliquots of ligand solution into the substrate solution. ITC is a standard method for determining the binding affinity of drug candidates to biological targets. From the heats of titration, it is possible to determine the enthalpy of binding and to construct a binding isotherm, from which the binding stoichiometry and affinity constant can be calculated.

7.5   Experimental Considerations/Best Practice

The ubiquity of heat[¶] is calorimetry's primary advantage and biggest drawback. Whilst IMC can be used to study almost any sample, it is often the case that the data are complex in shape because they may comprise contributions from several processes occurring simultaneously. It is also the case that calorimetric data are extremely susceptible to systematic errors because of the accidental measurement of one or more of a range of processes (such as solvent evaporation and/or condensation, erosion, side reactions and so on) that may occur concurrently with the study process(es). These errors become proportionately more significant as the heat output from the sample decreases. The majority of the effort in performing a calorimetric measurement should therefore be devoted to setting up the experiment to ensure that erroneous or unexpected powers have not been accidentally introduced as a corollary of poor experimental design or execution.

7.5.1   Interaction of the Sample with the Calorimeter Vessel

Consideration should be given to the potential for interaction between the sample and the calorimeter vessel. For example, acidic or basic solvents may corrode vessels made of a metal alloy (like Hastelloy), while steel interacts with halides and salts. Proteins and other biological materials tend to adhere to glass surfaces. In the case of an instrument with a fixed vessel, the material cannot be changed, but it may be possible to coat the inner surface (for instance with a silicone material like Repelcote) prior to use.

7.5.2   Selection of a Reference Material

A reference material should be inert and as closely matched to the sample material as possible, both in terms of the heat capacity and state. For example, if a solution phase reaction is to be studied then the reference material would be an identical volume of solvent. For solid

---

[¶]Heat does not come in different colours.[4]

samples, dried talc is usually appropriate. Isoperibolic calorimeters are not usually differential instruments, so no reference is needed.

7.5.3 Best Practice

Advice on best practice depends on the type of calorimetry measurement being made, but one factor is critical for nearly all isothermal measurements: the temperature of the sample (and its enclosure) must be identical to that of the instrument (the exception is isoperibolic solution calorimetry, where the sample and vessel are deliberately held 1–200 mK away from the instrument temperature). In most cases, this means that time must be given following loading of the sample for temperatures to equalise. Some instruments have an equilibration loading position, designed to allow heat exchange with the instrument before the sample is placed into the measuring position. If the sample is large and its temperature is significantly different from the instrument, then it should be pre-warmed in an incubator prior to loading (if placed directly into the calorimeter, there may be so much heat exchange that the instrument itself may change temperature). It is important to note that reaction may have commenced when the sample was loaded into the vessel, so the time between loading and the commencement of data capture should be recorded and must be appended to the start of the time column before any analysis is performed.

In the case of ampoule calorimetry, particular attention must be paid to ensuring a hermetic seal between the ampoule and the lid. If the seal is not air-tight then water or solvent in the sample will evaporate, a process that will manifest as an endothermic signal. It should also be ensured that ampoules are clean and dry prior to use. Rinsing an ampoule with a solvent and drying will significantly reduce its temperature, which will increase the equilibration time required to equalise to the temperature of the instrument.

There are many experimental factors to be considered when undertaking ITC measurements. Firstly, one species must be selected as the ligand and the other as the substrate. In principle, the choice is not important but there are several factors to consider:

- The volume of sample in the calorimeter is larger than the volume of solution injected
- The molar ratios of ligand and substrate must be selected carefully to ensure that binding saturation occurs during the experiment

 – The ligand solution is usually required to be quite concentrated
   (because only small aliquots of solution are injected to minimise
   thermal shocks and heat-capacity effects) so the solubility of the
   species is important

It is imperative that the same solvent is used to prepare both solutions, because the enthalpy of mixing/dilution of solvents, particularly if buffered, can be significant. Care is also needed if the
reaction produces a product that causes ionisation of a buffer, because the heat of ionisation can be considerable and should be corrected for. It is also important to ensure that where experiments are
conducted in different buffers, all solutions are of equivalent ionic
strength.

The number of injections is also important. Sufficient ligand solution must be injected to saturate the substrate (typically 15–25 injections, depending on the concentrations). A smaller number of
injections of larger volume will increase the measured heat per injection, but reduce the resolution of the binding isotherm. Also, the
time between the injections must be sufficient for the system to return to equilibrium (*i.e.* all of the reaction must be complete before
the next injection is made). For all ITC experiments, corrections must
be made for dilution of the components as the number of injections
increases.

For solution calorimetry, attention must be focused on preparation of the sample to be dissolved or dispersed. In particular, it is
essential that the mass of the sample to be studied is measured
accurately (the sample should have been stored in a desiccator prior
to loading so the mass is not affected by the presence of any water or
solvent and it may be necessary to correct for air buoyancy, which
can affect the mass by as much as 0.05% depending on sample
density). The sample should be freely soluble in the volume of
solvent used and, for isoperibolic instruments, dissolution should
be complete within 30 min.

Gas perfusion calorimetry requires consideration of the gas supply
as well as data interpretation. It is particularly important that:

 – The gas is dry (use of an in-line desiccant is recommended)
 – Solvent reservoirs are full and will not run dry during the course
   of the experiment
 – There are no gas leaks in the system (a flow meter on the outlet
   pipe provides confidence)
 – Solvents do not interact with the rubber o-rings

- External parts of the apparatus are maintained above the temperature of the calorimeter (to prevent condensation)
- Interaction with the sample is not rate-limited by the flow rate of the gas

Mass-flow controllers are usually used to ensure quantitative and proportional mixing of the gas lines. Placing a small amount of a saturated salt solution in the calorimeter vessel (for instance, a saturated solution of sodium chloride will maintain an RH of 75% at 25 °C. Nyqvist[5] provides a list of saturated salts and the relative humidities they will maintain) allows the programmed RH to be validated, Figure 7.5. If the RH of the gas is set below that of the salt solution, an endothermic power will be measured (as water evaporates to the atmosphere) and if the RH of the gas is above that of the salt solution, an exothermic power will be measured (as water condenses to the salt solution). When the RH of the gas equals that maintained by the salt solution, no power should be measured.

The moisture or solvent content of the sample must be reduced. This is most easily achieved by flowing dry gas over the sample during the initial phase of the experiment. The wetting response will be

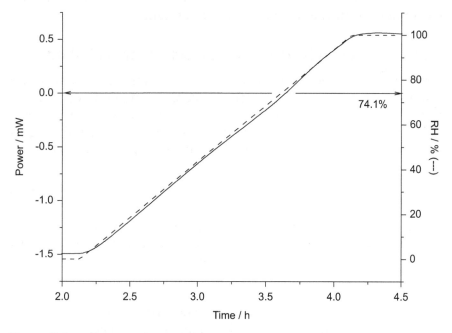

Figure 7.5 A graph of power *versus* time as the RH in the calorimeter ampoule was linearly increased. The ampoule held a small tube containing saturated NaCl solution (maintaining an RH of 75%).

affected by both the particle size distribution and the total mass of the sample. Reducing the particle size of the sample will increase the surface area to be wetted. The best option here is to sieve the sample prior to the experiment and select a particular particle size fraction for each measurement. Increasing the mass of sample will result in formation of a powder column and it is likely that material at the base of the column will not be exposed to vapour.

For flow calorimetry, the solution is pumped from an external reservoir through the calorimeter vessel. The reservoir must be thermostatted to the operating temperature of the calorimeter and it should be stirred. The peristaltic pump that circulates the solution must be calibrated for flow rate. Fast flow rates may cause noise in the power data and may also reduce the magnitude of the data by conducting heat away from the thermopiles.

7.6   Calibration

It is extremely important that any calorimeter is calibrated properly before use. Calibration is typically performed using the electrical substitution method, but chemical test and reference reactions are recommended to check calorimeter performance.

7.6.1   Calibration by the Electrical Substitution Method

Power-compensation instruments measure power directly and so no calibration constant (ε) is needed. Adiabatic and heat conduction calorimeters must be calibrated in order to determine the value of ε. Calibration is usually achieved by passing a known current through a resister located in or near the calorimeter vessel to produce a specific power:

$$\frac{dq}{dt} = I^2 R \tag{7.21}$$

where I is the current and R is the resistance. Alternatively, current can be passed through the resistor for a particular time. The area under the peak gives the heat produced:

$$q = I^2 R t \tag{7.22}$$

For heat conduction instruments, the heat produced by the resistor will generate a voltage from the thermopiles and hence:

$$\varepsilon = \frac{q_{calibration}}{\Delta V_{calibration}} \tag{7.23}$$

For adiabatic calorimeters, the heat produced by the resistor will generate a rise in temperature and hence:

$$\varepsilon = \frac{q_{calibration}}{\Delta T_{calibration}} \tag{7.24}$$

Most software packages will perform the calculation to determine the value of ε automatically.

7.6.2 Chemical Test and Reference Reactions

While the electrical substitution method is by far the most commonly used method of calibration, principally because of its ease of use and reproducibility, there are some potential problems with the procedure. The first is that the accuracy of the calibration is dependent entirely upon knowing the resistance of the heater (which will change with temperature) and the amount of current supplied. This can be mitigated to a certain degree by ensuring the heating resister is part of a Wheatstone bridge, as this allows determination of its resistance as a function of temperature. The second, and perhaps more significant, issue is that the processes of heat generation and dissipation from a resistance heater do not accurately mimic those that occur during a chemical reaction. This has led to much discussion and attempts to define standard chemical test reactions for isothermal calorimeters, a debate summarised in an IUPAC technical report.[6]

There are a number of requirements imposed upon a chemical test reaction: it must be chemically robust (*i.e.* it must give a consistent output over a range of conditions), it must have certified reaction parameters (such as a rate constant or an enthalpy of reaction/solution) and it must be repeatable from laboratory to laboratory and from instrument to instrument. If a chemical test reaction is run in a calorimeter, the user should, through appropriate analysis of the data, obtain values that agree with the literature parameters. This assures that (i) the instrument is functioning correctly and (ii) the operator is competent in the use of the technique.

It is notable here that chemical reactions are not generally intended to be used to calibrate instruments (that is still the role of the electrical heater); they are designed to give a reproducible and quantifiable heat output such that analysis in an electrically calibrated instrument returns a universally accepted set of values (be these rate constants, enthalpies *etc*). Proper selection of a test reaction depends upon both the type of instrument being used and the nature of the data being recorded. For instance, if the calorimeter is being used to

Table 7.1 Recommended test and reference reactions for various types of isothermal calorimetry.

Type of calorimetry	Test reaction	Reaction parameters	Notes	Reference
Ampoule or flow	Hydrolysis of triacetin	$\Delta_R H = -91.7 \pm 3$ kJ mol^{-1} $k = 2.8 \pm 0.1 \times 10^{-6}$ dm^3 mol^{-1} s^{-1}	Subjected to IUPAC inter- and intra-laboratory trial	7, 8
ITC	Neutralisation of NaOH with HCl (both 0.1M)	$\Delta_R H = -57.1$ kJ mol^{-1} at 25 °C	For enthalpy and/or volume validation	9
ITC	Binding of Ba^{2+} with 18-crown-6	$\Delta_R H = -31.42$ kJ mol^{-1}, $K_B = 5.9 \times 10^3$ mol dm^{-3}	For binding constant validation	10
ITC	Binding of 2$'$CMP with Rnase	$\Delta_R H = -50$ kJ mol^{-1}, $K_B = 1.2 \times 10^5$ mol dm^{-3}	For binding constant validation	11
Solution	Dissolution of KCl	17.584 ± 0.017 kJ mol^{-1}	NIST^a certified sample. Assumes final conc. of 0.111 mol kg^{-1}	12

^{a}National Institute for Standards and Technology.

measure the stability of a formulation over a period of days, then it is desirable to have a chemical test reaction that proceeds over several days and that gives a reasonable heat output; this ensures excellent reproducibility of measurement and allows both short-term (hours) and long-term (days and weeks) assessment of calorimetric performance. Conversely, if the calorimeter is being used in titration mode, then the test reaction should also involve a titration (a short-term test reaction), to allow validation of the heat per injection and the binding constant. Recommended test reactions for a number of isothermal calorimetric techniques are listed in Table 7.1.

7.7 Applications

The potential applications for IC in characterising materials are varied, and only brief summaries of the main areas of application are possible here. The simplest experiment involves placing a small (typically up to 100 mg) amount of sample (typically a solid) into an ampoule. If the sample undergoes a process, such as chemical degradation or change in physical form, then a power is seen. Unlike in the case of DSC experiments where endothermic events are often recorded, the power recorded by IC is almost always exothermic, because the reaction occurs spontaneously (if endothermic powers are seen, this is usually a result of water or solvent evaporation). Absence of a measurable power gives confidence that the sample is stable.

Quantitative assessment of stability can be performed in either the solid state (with control of RH) or in solution. If degradation follows first-order kinetics, analysis simply involves plotting ln(power) *versus* time. The slope of the resulting straight line gives the rate constant. The enthalpy of reaction is determined by calculating the area under the power-time curve (heat, Q, in J) and dividing it by the number of moles of compound that has reacted. Analysis is more complicated if other higher or mixed-order kinetics are involved. A comprehensive overview is given by Gaisford and O'Neill.[13]

Isothermal titration calorimetry is widely used to investigate the binding thermodynamics of substrates to biological ligands (and so to optimise lead compound design). Examples include its use to quantify the interaction of DNA with betaxolol[14] and characterise the solubilisation of simvastatin with a range of surfactants.[15]

In vivo applications of IC really focus on microbiology and biotechnology,[16] partly because the power from growing organisms is easily detected, even in complex or heterogeneous media. In addition, it is possible to study mixed bacterial populations or biofilms. Since the shape of bacterial growth curves is unique to each organism, IC can also be used to identify bacterial species.[17] It is thus possible to study the effect of antibiotics on bacteria or to investigate the growth and proliferation of bacteria on medical devices. For instance, Gaisford *et al.*[18] and Said *et al.*[19] used IC to quantify the effect of silver-containing wound dressings against *S. aureus* and *P. aeruginosa* while Said *et al.*[20] tested the efficacy of an anti-biofilm dressing. Similarly, human cell growth and interactions can be investigated directly.[21] Calorimetry is also used in the clinic to identify sepsis or urinary tract infections.[22]

7.8 Summary

While DSC is probably the most commonly used form of calorimetry, it is mainly used to investigate thermally-driven phase changes. IC is used to study time-dependent phase changes and so the two techniques form a powerful combination. The high sensitivity of many IC instruments means they can be used to monitor changes in very dilute solutions and the fact that heterogeneous or opaque materials can be investigated means IC is often applied to complex samples. The power-time data produced are usually very sensitive to small changes in the sample, which means the technique has great potential for batch-to-batch testing, but the ubiquitous nature of heat means data interpretation can be tricky and is often, at best, qualitative.

References

1. A. Lavoisier, P. S. Laplace, *de Histoire de l'Academie Royale des Sciences* 1780, 355.
2. L. D. Hansen, *Thermochim. Acta*, 2001, **371**, 19.
3. I. Wadsö, *LKB Inst. J.*, 1966, **13**, 33.
4. A. Cooper, C. M. Johnson, J. H. Lakey and M. Nöllmann, *Biophys. Chem.*, 2001, **93**, 215.
5. H. Nyqvist, *Int. J. Pharm. Technol. Prod. Manuf.*, 1993, **4**, 47.
6. I. Wadsö and R. N. Goldberg, *Pure Appl. Chem.*, 2001, **73**, 1625.
7. A.-T. Chen and I. Wadsö, *J. Biochem. Biophys. Methods*, 1982, **6**, 297.
8. A. E. Beezer, A. K. Hills, M. A. A. O'Neill, A. C. Morris, K. T. E. Kierstan, R. M. Deal, L. J. Waters, J. Hadgraft, J. C. Mitchell, J. A. Connor, J. E. Orchard, R. J. Willson, T. C. Hofelich, J. Beaudin, G. Wolf, F. Baitalow, S. Gaisford, R. A. Lane, G. Buckton, M. A. Phipps, R. A. Winneke, E. A. Schmitt, L. D. Hansen, D. O'Sullivan and M. K. Parmar, *Thermochim. Acta*, 2001, **380**, 13.
9. J. G. Stark and H. G. Wallace. *Chemistry data book*. 2nd edn, London: John Murray, 1982, p. 56.
10. L.-E. Briggner and I. Wadsö, *J. Biochem. Biophys. Methods*, 1991, **22**, 101.
11. M. Straume and E. Freire, *Anal. Biochem.*, 1992, **203**, 259.
12. G. A. Uriano, National Bureau of Standards Certificate. Standard Reference Material 1655, Potassium Chloride, KCl (cr) for Solution Calorimetry, 1981.
13. S. Gaisford and M. A. A. O'Neill, Pharmaceutical Isothermal Calorimetry, Informa Healthcare, 2007.

14. D. Z. Sun, X. Y. Xu, M. Liu, X. J. Sun, J. Y. Zhang, L. W. Li and Y. Y. Di, *Int. J. Pharm.*, 2010, **386**, 165.
15. R. Patel, G. Buckton and S. Gaisford, *Thermochim. Acta*, 2007, **456**, 106.
16. T. Krell, *Microbiol. Biotechnol.*, 2008, **1**, 126.
17. E. A. Boling and G. C. Blanchard, *Nature*, 1973, **241**, 472.
18. S. Gaisford, A. E. Beezer, A. H. Bishop, M. Walker and D. Parsons, *Int. J. Pharm.*, 2009, **366**, 111.
19. J. Said, C. C. Dodoo, M. Walker, D. Parsons, P. Stapleton, A. E. Beezer and S. Gaisford, *Int. J. Pharm.*, 2014, **462**, 123.
20. J. Said, M. Walker, D. Parsons, P. Stapleton, A. E. Beezer and S. Gaisford, *Int. J. Pharm.*, 2014, **474**, 177.
21. R. Santoro, O. Braissant, B. Muller, D. Wirz, A. U. Daniels, I. Martin and D. Wendt, *Biotechnol. Bioeng.*, 2011, **108**, 3019.
22. O. Braissant, D. Wirz, B. Gopfert and A. U. Daniels, *Sensors*, 2010, **10**, 9369.

8   Isothermal Reaction Calorimetry and Adiabatic Calorimetry

Ian Priestley

Syngenta Process Hazards Group, Huddersfield, UK
Email: ian.priestley@syngenta.com

8.1   Introduction and Principles

Isothermal reaction calorimetry (IRC) and adiabatic calorimetry (AC) are techniques that are commonly used within the chemical process industry for the development and safe scale-up of reactions. In order to operate any chemical process safely and economically, it is essential to have a comprehensive understanding of the effect that heat has on the reaction. This understanding can be split into two parts: what is intended to happen (the normal reaction) and what happens unintentionally (the abnormal, or thermal runaway). Although there are some overlaps, the normal reaction is typically characterised using isothermal reaction calorimetry and the abnormal reaction using adiabatic calorimetry.

For any given process, there is a minimum data set that is necessary to ensure safe operation and scale-up of a process, including:

(i) The thermal stability of reactants, mixtures and products
(ii) The enthalpy of reaction
(iii) The heat capacity of the system
(iv) The rate of heat production/removal

Principles of Thermal Analysis and Calorimetry: 2nd Edition
Edited by Simon Gaisford, Vicky Kett and Peter Haines
© The Royal Society of Chemistry 2016
Published by the Royal Society of Chemistry, www.rsc.org

  (v) The heat transfer properties of the system
  (vi) The reaction kinetics
 (vii) The quantity and rate of gas evolution

Although AC is primarily used to determine the thermal stability of materials, other thermal methods, often DSC and TGA, are also used, particularly during initial screening. Using either IRC or AC will allow data for items (ii) to (vi) to be obtained. Although a complete understanding of the gas-evolving properties of a chemical system (vii) is vitally important to ensure both economic and safe operation of a chemical process, the measurement of these properties is beyond the scope of this text.

For both IRC and AC, the important parameters being measured are the heat input/output of a reaction. These data are then used to perform a heat balance calculation on the reaction vessel, which gives an indication of the heat generated during the desired reaction and any subsequent heat release from accumulation of unreacted materials or undesirable side reactions.

8.2   Isothermal Reaction Calorimetry

Isothermal reaction calorimeters are basically small-scale versions of large chemical reactors. They allow chemical reactions to be carried out under controlled conditions whilst monitoring any changes in heat-output.

The heat obtained is often referred to as the 'heat of reaction', although this is not strictly correct as the overall heat output from a process will be made up of contributions from a number of sources, either chemical (*e.g.* reaction, dilution, crystallisation *etc.*) or physical (*e.g.* agitation *etc.*). Whilst it is possible to compensate for the physical effects either mathematically or by calibration, the chemical heat will cover a wide spectrum and, in the majority of cases, it is not possible to differentiate events easily or quickly unless hyphenated techniques using internal sensors (*e.g.* IR *etc.*) are used. Excellent reviews of isothermal reaction calorimetry are provided by Zogg *et al.*[1] and Cronin *et al.*[2]

Isothermal reaction calorimeters are available from a number of commercial suppliers, and, in addition, a number of research institutions and industrial companies have developed their own calorimeters that have not been commercialised. In order to avoid bias, only the basic theoretical principles are described.

Depending upon which particular model of reaction calorimeter is used, the usable reaction volume will be between *ca.* 40 mL and 2 L. Most experiments are carried out with a volume between 250 and 750 mL. Because of the comparatively large scale of the reaction, it is important to carry out suitable safety screening prior to operating on this scale.

There are a number of different principles of operation, some or all of which will be available to the user depending upon the type of calorimeter chosen, but all instruments maintain the environment around the calorimeter at a constant temperature. Exothermic or endothermic changes will produce a temperature increase or decrease in the reactor and the heat flow can be measured.

Although this type of calorimeter is simple to construct and maintain, it does have a number of disadvantages which make its general use problematic. The main disadvantages are:

(1) As the temperature is allowed to rise, the heat flow is non-linear, which makes obtaining kinetic data difficult.
(2) The non-linear response means that heat capacity measurements are problematic.
(3) It is difficult to alter the heating medium temperature whilst still obtaining meaningful data.

Isothermal reaction calorimeters control the reaction temperature over a very close range, typically ± 0.1 K. This is achieved in one of three ways:[3,4]

(1) Heat balance calorimetry in which a heating/cooling medium is circulated in a jacketed reactor and the temperature of the jacket fluid is varied so that the reaction temperature remains in the required range.
(2) Heat flow (or heat flux) calorimetry in which the heating/cooling medium is rapidly circulated in a jacketed reactor but minor reaction temperature changes are permitted.
(3) Power compensation calorimetry in which electrical power is provided to balance a cooling load.

Schematic representations of each type of calorimeter are given in Figure 8.1. Note that some isothermal reaction calorimeters have an adiabatic mode, which can control some or all of the heat output. The author urges caution if this mode is used. Due to the larger quantities of material involved in this type of experiment, if prior screening of the chemistry has not been carried out and evaluated appropriately, there is an increased risk of a thermal runaway and hence damage to the equipment and/or injury to personnel.

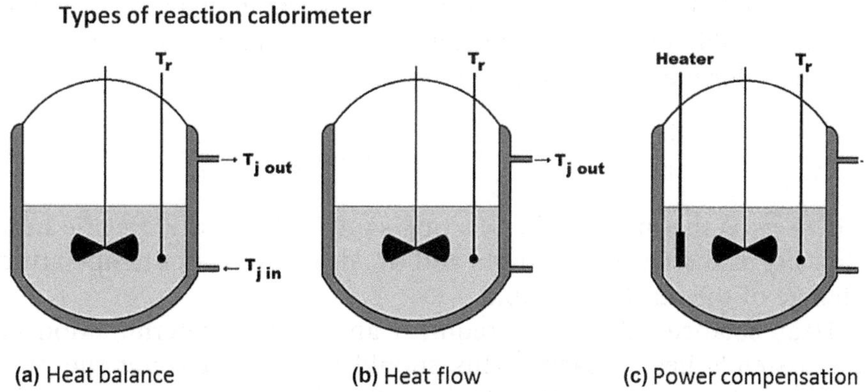

Types of reaction calorimeter

(a) Heat balance (b) Heat flow (c) Power compensation

Figure 8.1 Types of reaction calorimeter.

8.2.1 Heat Balance Calorimetry

Heat balance calorimeters mimic large-scale chemical reactors with a circulating heating/cooling medium maintaining the desired temperature. As a reaction occurs (either endothermic or exothermic), the temperature in the reactor will change, resulting in a change in the temperature of the circulating medium. By monitoring the inlet and outlet temperatures of the jacket, an adjustment in the medium temperature can be made, quickly bringing the reaction temperature under control. Although this is not totally isothermal, provided that the flow rate of the medium through the jacket is sufficiently rapid, effective temperature control to within a few tenths of a degree can easily be obtained. By monitoring the flow rate of the medium along with the inlet $(T_{j\,in})$ and outlet $(T_{j\,out})$ temperatures in the jacket, provided that the specific heat capacity of the circulating medium (C_{pj}) is known, it is possible to calculate the heat release rate from eqn (8.1).

$$\frac{dq}{dt} = m_j \cdot C_{pj} \cdot (T_{j\,out} - T_{j\,in}) \tag{8.1}$$

where dq/dt is the heat release rate (W) and m_j is the mass-flow of the circulating medium (g/s).

Eqn (8.1) represents an ideal situation. For most calorimeters, it is necessary to apply a correction factor to account for the power losses $\left(\frac{dq_{loss}}{dt}\right)$ in the system:

$$\frac{dq}{dt} = m_j \cdot C_{pj} \cdot (T_{j\,out} - T_{j\,in}) + \frac{dq_{loss}}{dt} \tag{8.2}$$

In addition to the heat loss term, it also may be necessary to make an adjustment for the fact that true isothermal conditions are never maintained (if the system was truly isothermal, then no reaction would be taking place—there needs to be some perturbation in order for the calorimeter to work). This leads to:

$$\frac{\mathrm{d}q}{\mathrm{d}t} = m_{\mathrm{j}}C_{\mathrm{p\,j}}(T_{\mathrm{j\,out}} - T_{\mathrm{j\,in}}) + m_{\mathrm{r}}C_{\mathrm{pr}}\frac{\mathrm{d}T_{\mathrm{r}}}{\mathrm{d}t} + \frac{\mathrm{d}q_{\mathrm{loss}}}{\mathrm{d}t} \qquad (8.3)$$

where m_{r} is the mass of the reacting sample, C_{pr} is the specific heat capacity of the reacting sample and $\mathrm{d}T_{\mathrm{r}}/\mathrm{d}t$ is the rate of temperature change of the reacting sample.

Heat balance calorimetry requires an accurate determination of the system's heat capacity. This should take into account the heat capacity of the reacting sample as well as the parts of the systems in contact with it (*i.e.*, the parts to which heat could be exchanged) such as the thermocouple, agitator, reactor wall *etc*. This determination is normally carried out once the system has reached equilibrium at the desired temperature just prior to commencing the reaction. The determination should also be repeated after the reaction is complete. The typical method is to use a calibrated electrical heater and apply a known power to the system for a set time period. A common source of error occurs when the calibration is carried out before the system has reached equilibrium.

This technique is very useful for reactions where there is a large volume change. It requires very high resolution of the ΔT measurement and can be used for reactions involving heating and cooling cycles provided that the specific heat capacity of the complete system (*i.e.*, reaction and the reactor) is known.

8.2.2 Heat Flow Calorimetry

In this variant of isothermal reaction calorimetry, the temperature of the jacket is kept constant by ensuring a very rapid flow of circulating medium. For very high flow rates, the temperature difference between the inlet and outlet of the circulating medium will tend towards zero, even if the heat output from the reaction is very high. This means that the heat flow signal is potentially much more accurate than the heat balance signal. For very high flow rates, the ability of the reactor to transfer heat from the reaction to the circulating medium is the important factor rather than the actual amount of coolant. The equation for heat release is therefore modified as follows:

$$\frac{\mathrm{d}q}{\mathrm{d}t} = UA(T_{\mathrm{r}} - T_{\mathrm{j}}) + m_{\mathrm{r}}C_{\mathrm{pr}}\frac{\mathrm{d}T_{\mathrm{r}}}{\mathrm{d}t} + \frac{\mathrm{d}q_{\mathrm{loss}}}{\mathrm{d}t} \qquad (8.4)$$

where U is the overall heat transfer coefficient, A is the heat transfer area, T_r is the temperature of the sample and T_j is the jacket temperature.

As most reactions involve a change in volume, this means that the heat transfer area and the overall heat transfer coefficient (OHTC) will change throughout the course of the reaction. Factors such as the speed of agitation (which may lead to the formation of a vortex) also change the heat transfer area. These variations potentially add a high degree of complexity to any measurements and subsequent calculations. Fortunately, calibration using an applied electrical power allows these variations to be considered.

$$\frac{dq_{calib}}{dt} = UA(T_r - T_j) + \frac{dq_{loss}}{dt} \tag{8.5}$$

where dq_{calib}/dt is the power input from the electrical heater.

The calibration needs to be carried out when the system is at equilibrium both before and after reaction and by calibrating the actual reaction mass being investigated to ensure the $\frac{dq_{loss}}{dt}$ term remains constant.

Changes in the physical nature of the reaction mass, *e.g.* viscosity or density, may have an adverse impact upon the heat transfer area and the OHTC of the calorimeter. For example, a sudden thickening of a batch during a reaction could prevent effective heat transfer meaning that any data derived from the experiment may not be an accurate representation of the actual situation. It is important that any user of this type of calorimeter has a good understanding of the principles and limitations of the technique and treats any data generated critically.

This type of calorimeter is most suitable for reactions in which there is only a small volume change. Where there is a significant volume change, multiple calibrations under steady state conditions are necessary to ensure accurate data are obtained. As with heat balance calorimeters, heating and cooling cycles can be accommodated provided that the specific heat capacity of the system is known.

8.2.3 Power Compensation Calorimetry

In power compensation calorimetry, a cooling load is applied to the reactor jacket. This cooling load is constant and is sufficient to maintain the reaction temperature at a fixed temperature, typically

10 to 20 K, lower than the desired reaction temperature. In order to overcome this temperature differential, a calibrated electric heater is used. Once steady state conditions are obtained, the addition of material can be started. As the reaction proceeds, in order to maintain 'isothermal' conditions for exothermic processes, the amount of electrical power supplied is reduced so that the desired temperature can be maintained (conversely, for endothermic processes, additional electrical power is supplied to maintain the reaction mass at the correct temperature). The electrical heat is therefore the inverse of the chemical heat in the process. In order for this to work effectively, a minor temperature perturbation of *ca.* 0.1 K is required.

Although good scientific practice dictates that the system is calibrated, the measurements can in fact be carried out without any direct calibration of individual reaction masses and the associated values of U and A. This technique is particularly useful for studying reactions with fast kinetics as the electrical heating element typically responds more rapidly than a circulating medium. It is, however, important to guard against localised heating from the electrical heater and this can be a problem with heat-sensitive or highly viscous materials, particularly if agitation is not effective.

There is a fourth type of reaction calorimeter that uses Peltier elements to control the reaction mass temperature by varying the power of the elements. It can work effectively to heat and cool the reaction mass. The heat flow is calculated based upon the electrical power and the ΔT over the Peltier elements. In theory, it would be possible to use Peltier elements in a calorimeter and a system partly based upon this method has been commercialised, but this utilises an additional sensor to monitor the heat rather than the Peltier function.

8.2.4 Calibration of Isothermal Reaction Calorimeters

For both the heat balance and heat flow methods, a calibration is typically applied once steady state conditions have been obtained but prior to the desired reaction taking place. A known electrical heat input is supplied to the system for a specific time such that the actual perturbation to the system can be recorded. This is shown schematically in Figure 8.2.

The ideal situation is shown on the left where the application of electrical power produces an immediate effect which is sustained throughout the event. In practice, calibrations are more likely to show the profile on the right in which the power signal takes some time to

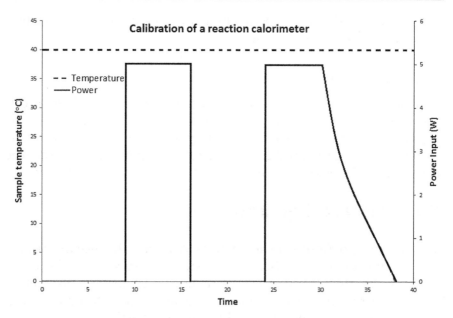

Figure 8.2 A typical calibration profile for an isothermal reaction calorimeter.

subside after the heat input is stopped. This is the response factor of the system and can vary depending upon a number of factors such as the equipment configuation and the materials being used. Calorimeter manufacturers take this response factor into account during the data analysis; this may require factory calibration of components and this can limit the flexibility and usability of the equipment.

Good practice is to carry out a similar calibration once the reaction is complete so that a comparison can be made and adjustments to the calculated enthalpy values made as appropriate. It is, however, important to ensure that the system has returned to steady state conditions before carrying out the calibration otherwise spurious values for the system heat capacity will be observed.

For power compensation calorimeters, direct calibration of the system is not always carried out as the technique relies upon using a calibrated electrical heater and the heat capacity of the system can be inferred from that data.

It is recommended that a cross check of the system heat capacity is made by comparing it to a calculated value (*e.g.*, group contribution method calculations) as this can easily highlight anomalous results, *e.g.*, slow residual reactions. In certain cricumstances, a standard reaction can be used for calibration; however, as this neccessitates carrying out additional experimental work, isothermal calorimeters are rarely calibrated in this manner.

8.2.5   Data Interpretation

Although the physical appearance of data obtained from different calorimeters may differ, Figure 8.3, the fundamental data and the way in which they are handled will be essentially the same. Simple integration of the power-time trace will give an indication of the total chemical heat output from the process from which the enthalpy can be calculated.

8.3   Adiabatic Calorimetry

The dictionary definition of adiabatic is 'A process that occurs without transfer of heat or matter between a system and its surroundings, *i.e.*, it obeys the first law of thermodynamics'. Adiabatic calorimeters operate on the principle of minimising heat losses to or from the surroundings, so cannot be classified as truly adiabatic.

The technique is primarily used in the study of chemical reaction hazards, in particular, runaway reactions, *i.e.*, reactions in which the heat output exceeds the cooling capability (either natural or forced).

A number of commercially available systems are available which simulate 'pseudo' adiabatic conditions by thermally insulating the sample, in most cases by placing the insulated sample inside an oven. The heat losses from the system can be minimised by using an oven temperature control system that monitors the temperature of the sample then adjusts the oven temperature to match the sample temperature. Alternatively, the temperature of the wall of the sample container is monitored and used as the reference temperature for temperature adjustment.

With this type of system, there is inevitably going to be some loss to the sample container but in order to simulate accurately a large scale chemical reactor or storage container, which typically lose less than 10% of the generated heat to the vessel itself, it is necessary to have a thermal inertia as close as possible to one.

The thermal inertia ratio is called the Phi (Φ) factor:

$$\Phi = 1 + \frac{C_{p\,\text{cell}}\, m_{\text{cell}}}{C_{p\,\text{sample}}\, m_{\text{sample}}} \tag{8.6}$$

where $C_{p\,\text{cell}}$ is the heat capacity of the cell, m_{cell} is the mass of the cell, $C_{p\,\text{sample}}$ is the heat capacity of the sample and m_{sample} is the mass of the sample.

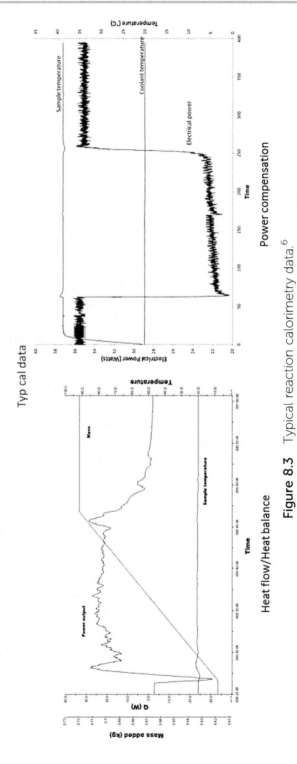

Figure 8.3 Typical reaction calorimetry data.[6]

Table 8.1 Influence of Phi factor on the increase in sample temperature (ΔT) and the predicted maximum temperature.

Phi factor	ΔT (K)	Max temperature (°C)
1.0	200	250
1.1	182	232
1.5	133	183
2.0	100	150
9.0	22	72

Under experimental conditions, the heat produced by the sample will increase both the temperature of the sample and that of the vessel. With a Phi factor of 1, the entire heat generated will be used to heat the sample and a worst case scenario can be simulated. As the Phi factor increases, the potential temperature rise is reduced (see Table 8.1). As the purpose of the experimental investigation is to determine what will happen in a worst case scenario, any reduction in the observed temperature rise as a result of losses to the equipment means that the worst case scenario has not been simulated correctly and the full thermal potential of the reaction will not be observed.

For example, assume the product from a reaction is thermally unstable on a 10 tonne scale from around 120 °C and from this temperature it will undergo a thermal runaway (*i.e.,* the rate of heat generation will exceed the rate of heat loss).

In a low phi factor simulation, the measured heat output from the process to form the material at 40 °C will result in a temperature rise of 90 K[†]. Therefore, if the reaction were carried out in an uncontrolled manner, sufficient heat would be generated to heat the batch to 130 °C and a thermal runaway would occur. In a high phi factor experiment, the measured heat output from the same process to form the material at 40 °C will result in a temperature rise of only 40 K. This would be sufficient to heat the batch to 80 °C, *i.e.*, 40 °C below the temperature from which a thermal runaway would occur. Therefore, in carrying out the experiment with a high phi factor, the actual plant situation would not be simulated and it would be incorrectly concluded that the reaction was safe.

A number of commercial systems compensate for a high Phi factor by data extrapolation and complex mathematics. For simple systems,

---

[†]Note here that although it is not mandatory, the Health and Safety Executive suggestion is to quote temperature rises in Kelvin because there have been a number of fatal incidents where temperature rises may have been mistaken for the reaction temperature.

this approach may be valid but for complex decomposition reactions, care must be exercised in adopting this approach and it is recommended that any high Phi factor experiments are repeated using an alternative technique, so that a comparison of the kinetic data can be made in order to determine if the data is usable.

8.3.1 Dewar Calorimetry[5]

This type of calorimeter is the simplest type of adiabatic calorimeter. They are normally used for investigating the thermal stability of materials and the heat output from batch and semi-batch processes. Dewar calorimeters use vacuum jacketed flasks (either glass or stainless steel). They provide accurate data on the rate and quantity of heat evolved in a reaction.

Typically, a calorimeter comprises a Dewar flask, which is placed inside an oven that is programmed to follow the temperature inside the Dewar, thus maintaining adiabatic conditions, Figure 8.4. The flask is fitted with a bung containing an agitator gland, thermocouple probe, gas off-take line and a means of addition of sample (liquid, gas, or solid). Maximum rates of self-heating that can be followed adiabatically are of the order of 30 K min^{-1}.

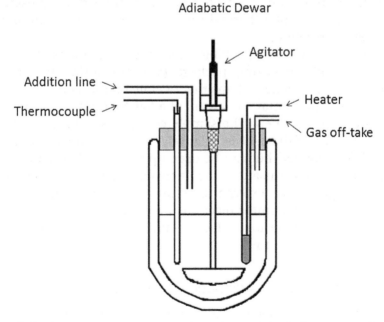

Figure 8.4   A schematic representation of an adiabatic Dewar calorimeter.

Dewar flasks have been shown to have cooling rates equivalent to large-scale plant vessels, which allows 'safe' small-scale simulation of large-scale processes to be carried out.[7,8]

8.3.2   Accelerating Rate Calorimetry (ARC)

The temperature range of the ARC is 0–500 °C with a pressure range from 0–150 bar.[9] It utilises a spherical sample holder with a volume of *ca.* 8–10 mL. The 'bombs' as they are called are usually made of titanium, Hastelloy or nickel alloys. The thermocouple is clipped to the outside of the 'bomb' and this can lead to problems with temperature measurement and exotherm detection particularly with viscous reaction masses or very fast reactions.

In order to prevent the 'bomb' bursting, it has thick walls, which means that it has a high thermal mass, hence the Phi factor tends to be on the high side with typical values being between *ca.* 1.5 and 9. Careful consideration of the data needs to be made in order to ensure that correct values for the adiabatic temperature rise are obtained. It is also advisable to use the ARC in conjunction with other techniques if kinetic analysis is to be performed.

Although the potential for catalysis from the metal bombs and the inability to either agitate or add materials once a test has commenced limit the usefulness of the technique, it is still extremely useful for detecting the start of a thermal runaway reaction and examining decomposition reactions.

Automatic pressure tracking adiabatic calorimetry (APTAC),[10] Phi-TEC[11] and vent-sizing package (VSP)[12] calorimetry are three commercially available systems that operate on the same principle as ARC. The sample containers are all of a low thermal mass so low Phi factors can be obtained. The sample containers are around 130 mL volume but are of themselves relatively weak and flimsy (similar to a soft drink container) but can be used up to relatively high pressures (200 bar) by compensating the internal pressure in the sample container against the external pressure in the equipment housing. They can typically monitor reactions up to around 500 °C and detect rates of self-heating as low as 0.02 K min^{-1} and can monitor self-heat rates of up to 400 K min^{-1}.

8.3.3   Calibration of Adiabatic Calorimeters

For adiabatic calorimeters, rather than the specific reaction under investigation, it is the system itself that is calibrated. A standard

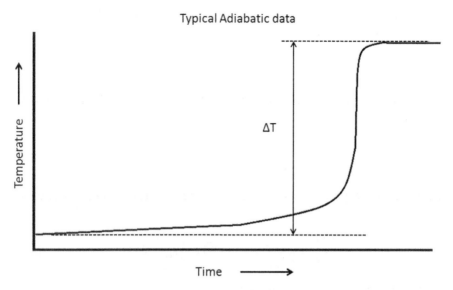

Figure 8.5 Typical adiabatic calorimetry data.

reaction that has known kinetic and thermodynamic parameters is normally used. The two most commonly used are the hydrolysis of acetic anhydride and the reaction between methanol and acetic anhydride. These two systems have the advantage that they are well characterised[13,14] and that they can also go to a thermal runaway thereby testing the equipment design as well as operation. Typical data are shown in Figure 8.5.

In order to determine the thermodynamic and kinetic parameters of a reaction, it may be necessary to apply a correction to the data to account for heat losses. For low Phi factor systems, *i.e.*, those that accurately simulate the large-scale situation, the adiabatic increase of temperature can directly be deduced from experimental data:

$$\Delta T_{ad} = T_{end} - T_{start} \tag{8.7}$$

where ΔT_{ad} is the temperature rise in the adiabatic calorimeter, T_{start} is the temperature of the sample at the start of the reaction and T_{end} is the temperature of the sample at the end of the reaction.

From this value, it is possible to estimate the total decomposition enthalpy (assuming a mean value for the specific heat of the reaction mixture) as follows:

$$\Delta H = m \cdot C_p \Delta T_{ad} \tag{8.8}$$

Obtaining kinetic information about the reaction is one of the important outcomes that can be obtained from adiabatic calorimetry.

Based upon the temperature rise, the extent of conversion or α can be determined as follows:

$$\alpha = \frac{(T - T_{start})}{\Delta T_{ad}} \tag{8.9}$$

Assuming that the reaction follows Arrhenius kinetics for an nth order reaction then:

$$\frac{dT}{dt} = Z \cdot e^{\left(\frac{-E}{RT}\right)} \cdot (1 - \alpha)^n \tag{8.10}$$

where Z is the pre-exponential factor and E is the activation energy.
  Hence,

$$\frac{dT}{dt} = \frac{Z}{(\Delta T_{ad})^{(n-1)}} \cdot e^{\left(\frac{-E}{RT}\right)} \cdot (T_{end} - T)^n \tag{8.11}$$

Eqn (8.11) can be re-written as:

$$\ln\left(\frac{dT}{dt}\right) = \ln\left(\frac{Z}{(\Delta T_{ad})^{(n-1)}}\right) - \frac{E}{RT} + n \cdot \ln(T_{end} - T) \tag{8.12}$$

This enables the activation energy and the order of reaction to be determined from the experimentally determined rates of temperature rise, because a plot of ln (self heat rate) against reciprocal absolute temperature will give a straight line of gradient $-E/R$. If a straight line is not obtained, it is evident that the simple kinetics assumed for the model are incorrect.
  For high Phi factor systems, an allowance has to be made for the heat capacity contribution from the reaction vessel as well as the sample itself. The formula for adiabatic temperature rise is therefore modified as follows:

$$\Delta T_{ad} = \Phi(T_{end} - T_{start}) \tag{8.13}$$

which in turn leads to eqn (8.12) being modified as follows:

$$\ln\left(\frac{dT}{dt}\right) = \ln\left(\frac{Z}{\Phi(T_{end} - T_{start})}\right) - \frac{E}{RT} + n \cdot \ln(T_{end} - T) \tag{8.14}$$

Initial evaluations of the data are often simplified by making a potentially worst case assumption that the reaction is concentration independent, *i.e.*, the reaction is zero order.[10]

$$\left(\frac{dT}{dt}\right) = \Phi\left(\frac{dT}{dt}\right)_{exptl} \tag{8.15}$$

Whilst this is generally applicable at the start of a slow reaction for most reactions, and in particular those in which there are complex decomposition processes occurring, this over-simplification can under-estimate the potential of a reaction.

This approach was modified by Huff [15] to include a correction for the rate of reaction that takes into account the Phi factor of the experiment:

$$\frac{\left(\dfrac{dT}{dt}\right)}{\left(\dfrac{dT}{dt}\right)_{\text{exptl}}} = \frac{1}{\Phi} \cdot \exp\left[-\frac{E}{R}\left(\frac{1}{T_{\text{ad}}} - \frac{1}{T_{\text{exptl}}}\right)\right] \tag{8.16}$$

Whilst this does give a more accurate representation of a single decomposition, for high Phi factor systems in which the complete exotherm is not observed, it will not provide the full decomposition profile.

A review by Wilcock *et al.*[16] considered various methods for correcting adiabatic data for Phi and concluded that the Huff model provided the most exact method of correcting for heat losses. The paper also indicated that physical parameters such as specific heat do not have a large impact upon the kinetics but slight changes in E/R or ln A can have a significant impact on the overall prediction.

8.4 Applications

The complimentary techniques of isothermal reaction calorimetry and adiabatic calorimetry were originally developed for use in the area of safe chemical process operation. The techniques allowed chemists to obtain an insight into potential thermal stability issues with materials, as well as an understanding of the changes in the heat output profile as a reaction proceeds. Due to the comparatively small scale, the experimentation can be carried out in a much safer environment than the 10's of m^3 often used in production, whilst still demonstrating the 'worst case' scenario.

It was soon recognised by process development chemists and chemical engineers that the availability of thermodynamic and kinetic data was not only useful in terms of safety but also provided valuable information for the scale-up of processes. One of the primary reasons for this was the fact that the small-scale laboratory calorimeters are able to closely mimic the large-scale plant reactors. The level of control of parameters such as temperature, addition rate, agitation *etc.* are significantly better than can be obtained from traditional laboratory glassware experiments. In many cases, the control is actually better than can be obtained on a full-scale chemical plant. In fact, reaction calorimeters have been used for the manufacture of small batches of commercially viable materials.

In addition to the role of adiabatic calorimetry in traditional process chemistry, there are an increasing number of applications in which they are being used to investigate 'new' technologies such as battery stability.[17]

8.5 Summary

Isothermal reaction calorimetry and adiabatic calorimetry are techniques that are commonly used within the process industries as an aid to the evaluation of process safety hazards and also to provide a better understanding of process parameters during the development phase. They are similar in principle to the isothermal instruments discussed in Chapter 7 but typically use larger sample sizes (200 mL to 2 L).

The techniques allow the generation of basic thermodynamic, kinetic and physicochemical property data of reaction mixtures and materials, which can be used to provide an understanding of the intended process chemistry and also the impact of process deviations.

As the experiments can be designed to mimic a large-scale plant operation, it is possible to use these comparatively small-scale techniques to simulate safely and accurately potentially hazardous processes such that safe and economic scale-up is possible.

References

1. A. Zogg *et al.*, *Thermochim. Acta*, 2004, **419**, 1.
2. J. L. Cronin, P. F. Nolan and P. E. Barton, IChemE Symp Series No 102, 1987, 113.
3. B. Grob *et al.*, *Thermochim. Acta*, 1987, **114**, 83.
4. J. Singh, Reaction Calorimetry for Process Development: Recent Advances, *Process Saf. Prog.*, 1997, **6**, 43.
5. N. Gibson, R. L. Rogers and T. K. Wright, *Hazards from Pressure Symposium Series* 1987, **102**, p. 61.
6. Y.-S. Duh *et al.*, *Thermochim. Acta*, 1996, **285**, 67.
7. R. L. Rogers, *Plant/Oper. Prog.*, 1989, **8**, 109.

8. T. K. Wright and R. L. Rogers, *IChemE Symp. Ser.*, 1986, **97**, 121–132.
9. D. I. Townsend and J. C. Tou, *Thermochim. Acta*, 1980, 37, 1.
10. W. Rogers *et al.*, *Thermochim. Acta*, 2004, **421**, 1.
11. J. Singh, International Symposium on Runaway reactions, Cambridge, Ma., AIChE/IChemE, 1989.
12. J. S. Sharkey *et al.*, *J. Loss Prev. Process Ind.*, 1994, 7, 413.
13. L. Friedel *et al.*, *J. Loss Prev. Process Ind.*, 1991, **4**, 110.
14. Y.-S. Duh *et al.*, *Thermochim. Acta*, 1996, **285**, 67.
15. J. E. Huff, *Plant/Oper. Prog.*, 1982, **4**, 211.
16. E. Wilcock and R. L. Rogers, *J. Loss Prev. Process Ind.*, 1997, **10**, 289.
17. T.-Y. Lu, *J. Therm. Anal. Calorim.*, 2013, **114**, 1083.

9 Thermomechanical, Dynamic Mechanical and Dielectric Methods

John C. Duncan[a] and Duncan M. Price[*b]

[a] Lacerta Technology, No. 3, The Courtyard, Main Street, Keyworth NG12 5AW, UK; [b] Edwards, Kenn Road, Clevedon BS21 6TH, UK
*Email: duncan.price@edwardsvacuum.com

9.1 Introduction and Principles

The effect of heat and cold upon the dimensions and mechanical properties of materials is of paramount importance to their use in an environment where they may encounter a wide variation in temperature through design or by accident. Plastics are processed at elevated temperatures so as to enable them to flow and be more amenable to fabrication, yet they must retain their form under normal usage. Food items are often heated and/or frozen before being consumed. Metals may be heat-treated to improve hardness and ceramics are fired so as to consolidate their final structure. This chapter covers various thermal methods that probe such behaviour and, due to common concepts, the effect of heat on the electrical properties of materials is also considered.

Principles of Thermal Analysis and Calorimetry: 2nd Edition
Edited by Simon Gaisford, Vicky Kett and Peter Haines
© The Royal Society of Chemistry 2016
Published by the Royal Society of Chemistry, www.rsc.org

9.1.1 Thermomechanical Analysis and Thermodilatometry

Thermomechanical analysis (TMA) is the study of a specimen's dimensions (length or volume) as a function of temperature whilst it is subjected to a constant mechanical stress. In this way, thermal expansion coefficients can be determined and changes in this property with temperature (and/or time) monitored. Many materials will deform under the applied stress at a particular temperature, which is often associated with the material melting or undergoing a glass–rubber transition. Alternatively, the specimen may possess residual stresses that have been "frozen-in" during preparation. On heating, dimensional changes may occur as a consequence of the relaxation of these stresses.

Stress (σ) is defined as the ratio of the mechanical force applied (F) divided by the area over which it acts (A):

$$\sigma = F/A \tag{9.1}$$

The stress is usually applied in compression or tension, but may also be applied in shear, torsion, or some other bending mode as shown in Figure 9.1. The units of stress are N m^{-2} or Pa.

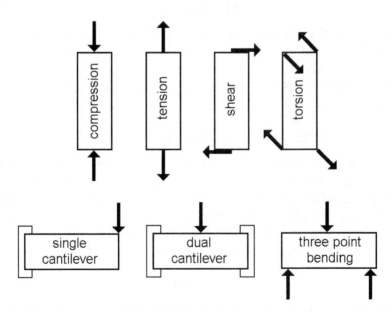

Figure 9.1 Common mechanical deformation modes: compression, tension, shear, torsion, bending (single cantilever, dual cantilever, three point bending).

If the applied stress is negligible then the technique becomes that of thermodilatometry. This technique is used to determine the coefficient of thermal expansion of the specimen from the relationship:

$$\alpha l_0 = dl/dT \tag{9.2}$$

where α is the coefficient of thermal expansion (ppm per °C), l_0 is the original sample length (m) and dl/dT is the rate of change of sample length with temperature (µm per °C).

9.1.2 Dynamic Mechanical Analysis

Dynamic mechanical analysis (DMA) is concerned with the measurement of the mechanical properties (mechanical modulus or stiffness and damping) of a specimen as a function of temperature. DMA is a sensitive probe of molecular mobility within materials and is most commonly used to measure the glass transition temperature and other transitions in macromolecules, or to follow changes in mechanical properties brought about by chemical reactions.

For this type of measurement, the specimen is subjected to an oscillating stress, usually following a sinusoidal waveform:

$$\sigma(t) = \sigma_{max}\sin \omega t \tag{9.3}$$

Where $\sigma(t)$ is the stress at time t, σ_{max} is the maximum stress and ω is the angular frequency of oscillation. Note that $\omega = 2\pi f$ where f is the frequency in Hertz.

The applied stress produces a corresponding deformation or strain (ε) defined by:

$$\varepsilon = (\text{change in dimension})/(\text{original dimension}) = \Delta l/l_0 \tag{9.4}$$

The strain is measured according to how the stress is applied (*e.g.* compression, tension, bending, shear *etc.*). Strain is dimensionless, but often expressed in %. The strain above would be for tensile or compressional deformation. Other factors can be seen in Table 9.1.

For an elastic material, Hooke's law applies and the strain is proportional to the applied stress according to the relationship:

$$E = d\sigma/d\varepsilon \tag{9.5}$$

where E is the elastic, or Young's modulus, with units of Nm^{-2} or Pa. Such measurements are normally carried out in tension or bending; when the sample is a soft material or a liquid then measurements are

Table 9.1 Geometry factors and strain factors.[a,b]

Geometry	Modulus	Geometry factor, k	Strain factor, f
Tension and compression	E^*	A/l	$1/l$
Single cantilever	E^*	$w(t/l)^3$	$3t/l^2$
Dual cantilever	E^*	$2w(t/l)^3$	$3t/l^2$
3 Point bending	E^*	$w/2.(t/l)^3$	$3/2.t/l^2$
Simple shear (double sample)	G^*	$2A/t$	$1/t$

[a]Where stiffness, $S^* = \sigma_{max}/y$ and where modulus, E^* or $G^* =$ stiffness, S^*/k and dynamic strain, ε or $\gamma = y.f$.
[b]Where y is the dynamic displacement amplitude, w is the width, t is the thickness, l is the free length (or sample height for compression) and A is the sample cross-sectional area. All size dimensions in metres.

normally carried out in shear mode, thus a shear modulus (G) is measured. The two moduli are related to one another by:

$$G = E/(2 + 2v) \tag{9.6}$$

where v is known as the Poisson's ratio of the material. This normally lies between 0 and 0.5 for most materials and represents a measure of the distortion that occurs (*i.e.* the reduction in breadth accompanying an increase in length) during testing.

If the material is viscous, Newton's law holds. The specimen possesses a resistance to deformation or viscosity, η proportional to the rate of application of strain, *i.e.*:

$$\eta = d\sigma/(d\varepsilon/dt) \tag{9.7}$$

The units of viscosity are Pa s.

A coil spring is an example of a perfectly elastic material in which all of the energy of deformation is stored and can be recovered by releasing the stress. Conversely, a perfectly viscous material is exemplified by a dashpot, which resists extension with a force proportional to the strain rate but affords no restoring force once extended, all of the deformation energy being dissipated as heat during the loading process. In reality, most materials exhibit behaviour intermediate between springs and dashpots—viscoelasticity.

If, as in the case of DMA, a sinusoidal oscillating stress is applied to a specimen, a corresponding oscillating strain will be produced. Unless the material is perfectly elastic, the measured strain will lag behind the applied stress by a phase difference (δ) shown in Figure 9.2. The ratio of peak stress to peak strain gives the complex

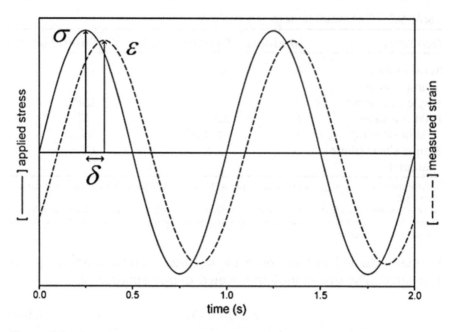

Figure 9.2 Relationship between applied stress (σ) and strain (ε) during a dynamic mechanical test. δ is the phase difference.

modulus (E^*), which comprises an in-phase component or storage modulus (E') and a 90° out-of-phase (quadrature) component or loss modulus (E'').

The storage modulus, being in-phase with the applied stress, represents the elastic component of the material's behaviour, whereas the loss modulus, deriving from the condition at which dε/dt is a maximum, corresponds to the viscous nature of the material. The ratio between the loss and storage moduli (E''/E') gives the useful quantity known as the mechanical damping factor (tan δ), which is a measure of the amount of deformational energy that is dissipated as heat during each cycle. The relationship between these quantities can be illustrated by means of an Argand diagram, commonly used to visualise complex numbers, which shows that the complex modulus is a vector quantity characterised by magnitude (E^*) and angle (δ) as shown in Figure 9.3. E' and E'' represent the real and imaginary components of this vector thus:

$$E^* = E' + iE'' = \sqrt{(E'^2 + E''^2)} \tag{9.8}$$

So that:

$$E' = E^* \cos \delta \tag{9.9}$$

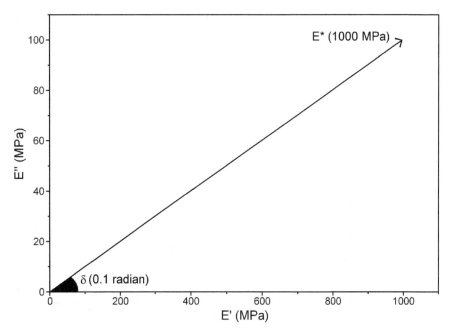

Figure 9.3 Argand diagram to illustrate the relationship between complex modulus (E^*) and its components.

and

$$E'' = E^* \sin \delta \tag{9.10}$$

9.1.3 Dielectric Techniques

In a manner analogous to TMA and DMA, a specimen can be subjected to a constant or oscillating electric field rather than a mechanical stress during measurements. Dipoles in the material will attempt to orient with the electric field, while ions, often present as impurities, will move toward the electrode of opposite polarity. The resulting current flow is similar in nature to the deformation brought about by mechanical tests and represents a measure of the freedom of charge carriers to respond to the applied field (Figure 9.4). The specimen is usually presented as a thin film between two metal electrodes so as to form a parallel plate capacitor. Two types of test can be performed:

9.1.3.1 Thermally Stimulated Current (TSC) Analysis

In this technique, the sample is subjected to a constant electric field and the current that flows through the sample is measured as a

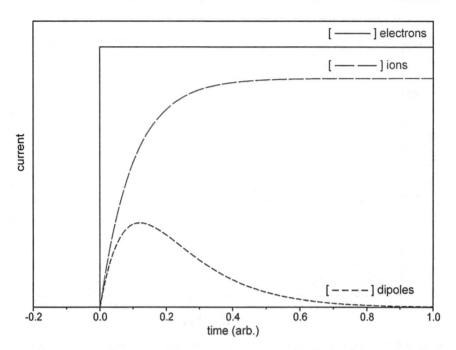

Figure 9.4   Response of electrons, ions and dipoles to the application of an electric field.

function of temperature. Often, the sample is heated to a high temperature at which point the static electric field is applied and then quenched to a low temperature. This process aligns dipoles within the specimen in much the same way that drawing a material under a mechanical stress would bring about orientation of molecules in the sample. The polarisation field is then switched off, and the sample is re-heated whilst the current resulting from the relaxation of the induced dipoles to the initial orientation or equilibrium position is monitored.

The simple experiment described affects all dipoles in the specimen that are able to realign with the applied electric field at the chosen polarisation temperature. The result of such a treatment is known as the "global" thermally stimulated current (TSC) signal. A more elaborate experiment is known as relaxation map analysis (RMA), which involves a series of experiments known as "thermal windowing" or "fractional polarisation" in which the global response is experimentally deconvoluted into specific groups of dipoles. This then allows a detailed investigation of the global relaxation process. The aim of the experiment is to choose thermal windowing conditions that isolate a single Debye relaxation process. In reality, this is never achieved, even over very narrow temperature windows as each process

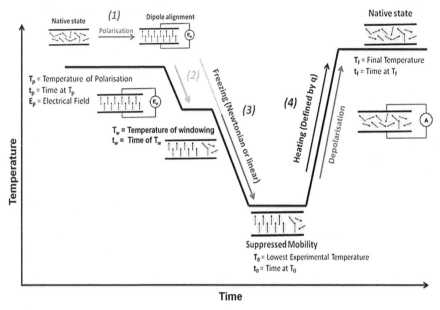

Figure 9.5 Schematic diagram of thermal windowing employed in TSC.

is characterised by a distribution of relaxation times, which must be determined and analysed to extract thermokinetic data.

An exaggerated schematic diagram of the thermal windowing procedure is shown in Figure 9.5. The steps in the experiment are:

(1) Sample is polarised at T_p for a time t_p.
(2) Sample is cooled to temperature just below T_p (T_d) *i.e.* 1 °C (T_w) below T_p. The external electrical field is short circuited and held isothermal for a time t_w allowing faster relaxing dipoles to relax back to equilibrium at that temperature.
(3) The remaining oriented dipoles are frozen in by cooling the sample to T_0.
(4) Sample is heated and the frozen-in dipoles relax back to equilibrium.

An example of this procedure will be considered later in this chapter.

9.1.3.2 Dielectric Thermal Analysis (DETA)

In this technique, the sample is subjected to an oscillating sinusoidal electric field. The applied voltage (V) produces a polarisation within the sample and causes a small current (I) to flow, which leads the

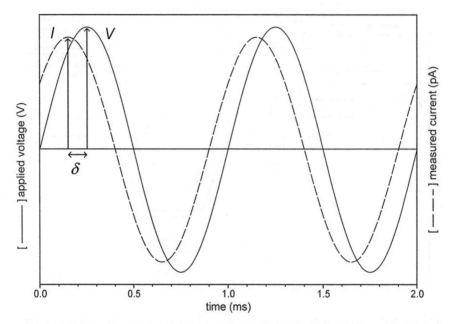

Figure 9.6 Relationship between voltage and current in a capacitor (c.f. Figure 9.2).

electric field by a phase difference (δ) (Figure 9.6). Two fundamental electrical characteristics, conductance and capacitance, are determined from measurements of the amplitude of the voltage, current and δ. These are used to determine the admittance of the sample (**Y**) given by:

$$\mathbf{Y} = I/V \tag{9.11}$$

Y is a vector quantity, like E^* discussed earlier, and is characterised by its magnitude $|Y|$ and direction δ.

Capacitance (C) is the ability to store electrical charge and is given by:

$$C = |Y| \sin \delta/\omega \tag{9.12}$$

Conductance (G) is the ability to transfer electric charge and is given by:

$$G = |Y| \cos \delta \tag{9.13}$$

More usually, data are presented in terms of the relative permittivity (ε') and dielectric loss factor (ε''); these are related to capacitance and conductivity by:

$$\varepsilon' = C/(\varepsilon_0 \cdot A/D) \tag{9.14}$$

and

$$\varepsilon'' = G/(\omega \cdot \varepsilon_0 \cdot A/D) \tag{9.15}$$

Where ε_0 is the permittivity of free space (8.86×10^{-12} F m^{-1}) and A/D (in m), is the ratio of electrode area (A) to plate separation or sample thickness, D for a parallel plate capacitor. More generally, A/D is a geometric factor that is found by determining the properties of the measuring cell in the absence of a sample. ε' and ε'' are dimensionless quantities.

The ratio $\varepsilon''/\varepsilon'$ is the amount of energy dissipated per cycle divided by the amount of energy stored per cycle and is known as the dielectric loss tangent or dissipation factor ($\tan \delta$).

9.2 Instrumentation

9.2.1 Thermomechanical Analysis

A schematic diagram of a typical instrument is shown in Figure 9.7. The sample is placed in a temperature-controlled environment with a thermocouple or other temperature-sensing device, such as a platinum resistance thermometer, placed in close proximity. The facility to circulate a cryogenic coolant such as cold nitrogen gas from a Dewar vessel of liquid nitrogen is useful for sub-ambient measurements. The atmosphere around the sample is usually controlled by purging the oven with dry air or nitrogen. Because of the much larger thermal mass of the sample and oven compared with a differential scanning calorimeter or a thermobalance, the heating and cooling rates employed are usually much slower for TMA. A rate of 5 °C min^{-1} is usually the maximum recommended value for good temperature equilibration across the specimen. Even this rate can be a problem for some samples where appreciable temperature gradients can exist between the middle and ends of the sample particularly around the test fixtures—which can represent a significant heat sink.

For compression measurements (as illustrated), a flat-ended probe is rested on the top surface of the sample and a static force is applied by means of an electromagnetic motor similar in principle to the coil of a loudspeaker. Some form of proximity sensor measures the movement of the probe. This is usually achieved by using a linear variable differential transformer (LVDT), which consists of two coils of wire that form an electrical transformer when fed by an AC current. The core of the transformer is attached to the probe assembly and the

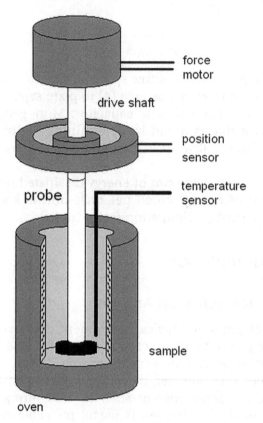

Figure 9.7 Schematic diagram of a thermomechanical analyser.

coupling between the windings of the transformer is dependent upon the displacement of the probe. Other transducers, such as capacitance sensors (which depend on the proximity of two plates—one fixed the other moving) or optical encoders, are used in certain instruments.

Most commercial instruments are supplied with a variety of probes for different applications (Figure 9.8). A probe with a flat contact area is commonly used for thermal expansion measurements where it is important to distribute the applied load over a wide area. Probes with sharp points or round-ended probes are employed for penetration measurements so as to determine the sample's softening temperature.[1] Films and fibres, which are not self-supporting, can be measured in extension by clamping their free ends between two grips and applying sufficient tension to the specimen to prevent the sample buckling. Volumetric expansion can be determined using a piston and cylinder arrangement with the sample surrounded by an inert

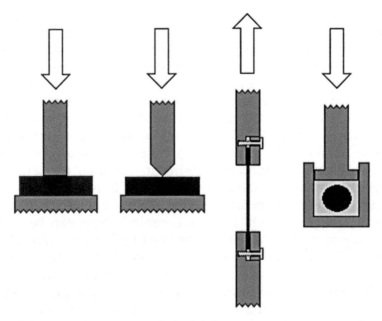

Figure 9.8 TMA probe types (left–right): compression, penetration, tension, volumetric.

packing material such as alumina powder or silicone oil. Other geometries include a three point bending arrangement, which can be used to determine a material's heat distortion temperature.[2]

The equipment must be calibrated before use. Instrument manufacturers, as well as various standardisation agencies, usually provide recommended procedures. Temperature calibration can be carried out by preparing a sample comprising a number of metal melting point standards, such as those used for differential scanning calorimetry, sandwiched between steel or ceramic discs. As the temperature of the furnace is raised, each metal melts and flows resulting in a step-wise change in height of the stack.[3] Force calibration of a thermomechanical analyser is often performed by balancing the force generated by the electromagnetic motor against a certified weight added to the drive train.[4] Length calibration can be more difficult to carry out. A common check on the performance of the instrument is to measure the thermal expansion of a material whose values are accurately known (such as aluminium or copper).[5,6]

9.2.2 Dynamic Mechanical Analysis

The distinction between a thermomechanical analyser and a dynamic mechanical analyser is blurred nowadays since many instruments can perform TMA-type experiments. The configuration of a DMA is essentially the same as the TMA shown in Figure 9.5 with the addition of extra electronics to apply an oscillating load and the ability to resolve the resulting specimen deformation into in-phase and out-of-phase components so as to determine E', E'' and $\tan \delta$. The facility for sub-ambient operation is more common on a DMA than a TMA. The same recommendations about modest rates of temperature change are even more important for the larger samples used in DMA. Stepwise-isothermal measurements are often carried out for multiple frequency operation. In this experiment, the oven temperature is changed in small increments and the sample is allowed to reach thermal equilibrium before the measurements are made. The frequency range over which the mechanical stress can be applied commonly covers 0.01 to 100 Hz. The lower limit is determined by the amount of time that it takes to cover enough cycles to attain reasonable resolution of $\tan \delta$ (100 s for one measurement at 0.01 Hz—though normally, some form of data averaging is applied meaning that a measurement at this frequency can take a minute or more). The upper limit is usually determined by the mechanical properties of the drive system and clamps.

Different clamping geometries are used to accommodate particular specimens (Figure 9.9). Single or dual cantilever bending modes are the most common for materials that can be formed into bars. Shear measurements are used for soft, thick samples. Films and fibres are usually mounted in tension with loading arranged so that the sample is always in tension.

Linear displacement DMAs are considered here. Rotational instruments producing torsional deformation can perform dynamic shear measurements on bar and plate samples and generate data similar to linear DMAs. Such instruments are called oscillatory or dynamic rheometers and are most commonly used in the study of melts and liquids. Rheometers have one plate or bar clamp fixed while the other rotates back and forth so as to subject the sample to a shearing motion. All of the normal dynamic parameters are calculated, shear storage modulus (G') and shear loss modulus (G'') and $\tan \delta$ from the ratio G''/G'. Other geometries such as concentric cylinders or cone and plate are often used to optimize the instrument's measurement range according to the viscosity of the sample.

The method of calibration of DMAs varies from instrument to instrument and it is essential to follow the manufacturer's recommendations. Temperature calibration can sometimes be done as for TMAs since many instruments can operate in this mode. Load or force

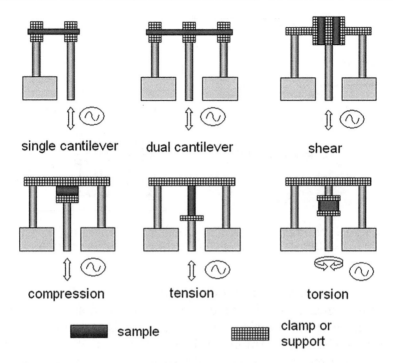

single cantilever   dual cantilever   shear

compression   tension   torsion

sample   clamp or support

Figure 9.9 Common clamping geometries for dynamic mechanical analysis (*c.f.* Figure 9.1).

calibration is often carried out using weights. It is difficult to achieve the same degree of accuracy and precision in modulus measurements from a DMA as might be obtained by using an extensometer without taking great care to eliminate clamping effects and the influence of instrument compliance (which can be estimated by measuring the stiffness of a steel beam). Extensometers are much bigger instruments and the size of test specimens is correspondingly larger. Additionally, they often only operate at room temperature. For many applications, the user is, however, mainly interested in the temperatures at which changes in mechanical properties occur and the relative value of a material's properties over a broad range of temperatures.

9.2.3 Dielectric Techniques

A schematic diagram of a typical instrument is shown in Figure 9.10. The sample is presented as a thin film, typically no more than 1 or 2 mm thick, between two parallel plates so as to form a simple

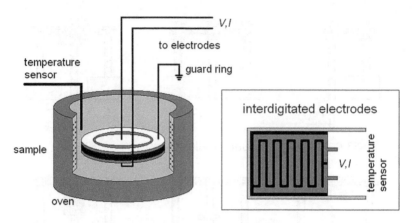

Figure 9.10 Schematic diagram of a dielectric thermal analysis instrument (inset shows single-surface interdigitated electrode).

electrical capacitor. A grounded electrode surrounding one plate, known as a guard ring, is sometimes incorporated so as to improve performance by minimising stray electric fields. A thermocouple or platinum resistance thermometer is placed in contact with one of the plates (sometimes one on each plate) so as to measure the specimen's temperature. For specialised applications, such as remote sensing of large components, an interdigitated electrode is used, shown inset into Figure 9.10. These employ a pair of interlocking comb-like electrodes and often incorporate a temperature sensor (resistance thermometer). These can be embedded in structures such as a thermosetting polymer composite and the dielectric properties of the material monitored while it is cured in an autoclave.

A usual part of the calibration protocol for DETA is to measure the dielectric properties of the empty dielectric cell so as to take into account stray capacitances arising from the leads, which must be of coaxial construction. Temperature calibration can be done by measuring the melting transition of a crystalline low molecular weight organic crystal such as benzoic acid placed between the electrodes.

9.3 Typical Experiments

This section discusses the most common types of experiments performed using TMA, DMA, TSC and DETA by way of introduction to some of the more advanced applications described later.

9.3.1 Thermomechanical Analysis

Thermomechanical measurements can be carried out on a wide range of solid samples. The most usual mode of measurement is either in compression (for self-supporting samples) or tension (for thin films and fibres). Some materials exhibit anisotropic behaviour (particularly films or crystals) in that changes in dimensions will differ depending upon which axis the measurements are performed.

9.3.1.1 Thermal Expansion Measurements and Softening Temperatures

A plot of the change in length of a sample of polytetrafluoroethylene (PTFE) is shown in Figure 9.11. The sample was heated at 5 °C min^{-1} under a static air atmosphere with a negligible load on the flat-ended probe. As the temperature increases, the specimen expands until the melting point of the polymer is reached around 326 °C. This is accompanied by a small amount of shrinkage. The glass–rubber transition of PTFE is not observed clearly on this plot because the amorphous content is very small, but closer inspection of the curve shows that there is a very small change in the slope of the plot around

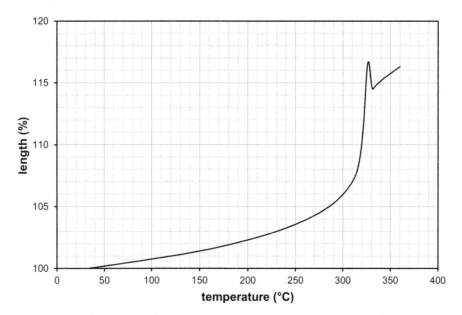

Figure 9.11 Plot of change in length for a sample of polytetrafluoroethylene under a flat-ended probe of 0.92 mm diameter with an applied load of 1 N. Heating rate 5 °C min^{-1} under nitrogen.

160 °C due to this process. At this temperature, the polymer chains acquire additional degrees of mobility, which is seen as an increase in thermal expansion coefficient.

Measurements of thermal expansion coefficients are useful in assessing the compatibility of different materials for fabrication into components. Mismatches in behaviour can cause stresses to build up when temperature changes occur, resulting in eventual weakening and failure of the structure. Many crystalline materials can exist in a number of polymorphic forms, which are stable at different temperatures. The transition between crystal structures is usually accompanied by a change in density and thermal expansion coefficient, which can be detected by TMA.

Supporting information from differential scanning calorimetry is often useful in interpreting information from TMA, particularly when softening point determinations are made, since loss of mechanical integrity can occur due to melting, which gives an endothermic peak in DSC or a glass–rubber transition, which causes a step change in heat capacity.

9.3.1.2   Force Ramp Experiments

Some TMAs are able to change the force on the sample during measurement so as generate force-displacement curves in a manner similar to a conventional extensometer with the additional advantage of good control of specimen temperature. As the stress on the specimen is increased, the material may creep under the applied load. When the force is removed, the sample may attempt to recover its original dimensions (stress relaxation). The stiffness of the material can be deduced from the relationship between the applied force and any change in dimensions that result. Such tests are useful is assessing the resilience of materials such as rubber gaskets, O-rings and the like. An example of this is shown in Figure 9.12 for a series of flexible sealing materials which were immersed in a mixed acid bath to simulate exposure to process chemicals. The tests were carried out at 50 °C, so the experiment does not really fall under the term "thermal analysis", but illustrates the use of a thermomechanical analyser to generate stiffness data relevant to an accelerated ageing study. In this example, it can be seen that the silicone material quickly degrades but, although the fluoropolymer material exhibits an initial loss in stiffness, it does not get any worse; whereas the performance of the diene polymer gradually decreases to a point where it becomes less resilient than the fluoropolymer.

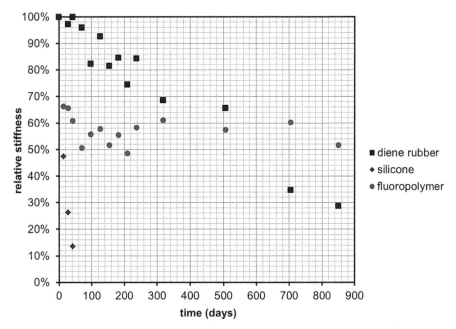

Figure 9.12 Stiffness of flexible seals during accelerated ageing in mixed acids.

9.3.2 Dynamic Mechanical Analysis

Dynamic mechanical analysis is routinely used to investigate the morphology of polymers, composites and other materials. The technique can be particularly sensitive to low energy transitions that are not readily observed by differential scanning calorimetry. Many of these processes are time-dependent and by using a range of mechanical deformation frequencies, the kinetic nature of these processes can be investigated.

9.3.2.1 Single Frequency Temperature Scans

The most common DMA experiment is simply to measure the storage modulus (E') and mechanical damping factor $(\tan \delta)$ against temperature at a single oscillation frequency. Figure 9.13 shows these properties for a stick of chewing gum, heated from $-50\,^{\circ}\text{C}$ to $+50\,^{\circ}\text{C}$ at a heating rate of $2\,^{\circ}\text{C min}^{-1}$ and a mechanical oscillation frequency of 1 Hz. The stiffness of the material decreases by over four orders of magnitude over this temperature range and it exhibits a broad damping peak centred roughly about $30\,^{\circ}\text{C}$. This mechanical

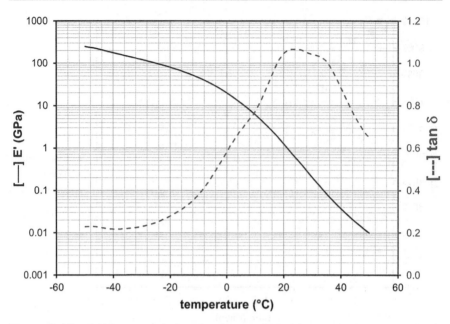

Figure 9.13   DMA curve of chewing gum at 1 Hz (single cantilever bending).

behaviour does not come about by accident: in order to be an appealing substance to masticate, the gum should have a leathery texture, exactly as observed within the glass transition region of most polymers. This is no coincidence as chewing gum is formulated to be most "chewy" at body temperature and the shape of the tan δ curve arises from different components (such as styrene–butadiene rubber, isobutylene/isoprene copolymer and petroleum wax) within the gum. This simple DMA thermal scan yields the necessary mechanical properties relevant to the application and the technique has been widely used for the characterisation of many other foods.[7–9]

9.3.2.2   Relaxation Processes

An example of another thermal scan can be seen in Figure 9.14. This shows a poly(ethylene terephthalate) (PET) sample tested from $-150\,^\circ$C to its melting point. The sample was tested as a single cantilever at two applied frequencies of 1 and 10 Hz. For DMA and DETA work, the glass transition is often called the alpha (α) transition and all lower temperature transitions are given corresponding Greek symbols beta (β), gamma (γ) *etc*. The first feature to note is the frequency dependence of the α transition (the glass transition) and β peaks. This confirms that they are relaxation processes. The β

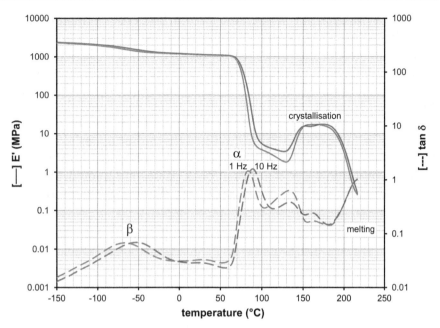

Figure 9.14 Relaxation processes in PET at 1 and 10 Hz (single cantilever bending).

transition in the polymer is due to the motion of short lengths of the polymer backbone rather than the large-scale increase in mobility that accompanies the glass–rubber transition. It should be noted that β transitions are frequently due to side chain motion. Under such circumstances, the effect is smaller than that observed here, since this process has its origin in the main backbone. It is very difficult to measure this type of behaviour by DSC, but the size and position of these transitions are often very important for a polymer's impact properties since they provide a means of dissipating mechanical energy as heat.[10]

The peak at ~90 °C is the glass transition or α process. Again, this is frequency dependent. Peaks at higher temperatures are due to crystallisation and finally the melting is observed which is independent of frequency, confirming the first-order nature of the melting process.

9.3.2.3 Defining the Glass Transition Temperature

Unlike the melting point, which is a first-order process, the glass transition process is second-order. This means that there is no instantaneous change in enthalpy or volume as the material is heated through T_g unlike that associated with a first-order transition such as melting. In a TMA experiment, for example, there is only a change in the rate of expansion. In a DSC experiment, the glass transition temperature is found to be heating rate dependent and similarly, in a DMA experiment, it is dependent upon the frequency of the applied load, lower frequencies yielding lower T_g's (see Figure 9.14). A further complication is that many of the DMA parameters can be used as a measure of T_g. Figure 9.15 shows a filled thermoset material and all of the values listed in Table 9.2 can be legitimately quoted for T_g.

Most commonly, the E'' or tan δ peak values are quoted. ISO 6721 Part 11 for the determination of T_g of composites states that the

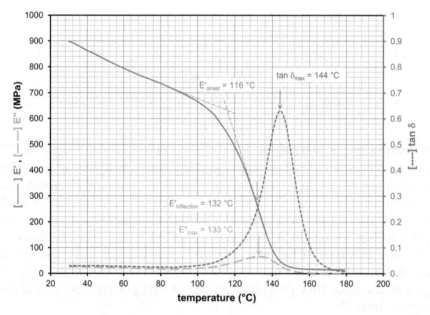

Figure 9.15 T_g definitions for the glass–rubber transition.

Table 9.2 T_g values measured in a single cantilever bending at 1 Hz.

Parameter	Feature	Value (°C)
E'	Linear onset	116
E'	Logarithmic onset	128
E'	Inflection point of step	132
E''	Peak	133
Tan δ	Peak	144

inflexion point in E' (as measured by its derivative) should be quoted as T_g, but other values may be given as well. It is also important to note the geometry used and the frequency of the applied load. For a typical amorphous T_g around $100\,^\circ$C, the peak positions for T_g shift approximately $7\,^\circ$C per decade of frequency. See Section 9.4.2.1 for further comments on the use of E'' or $\tan\delta$ peak values in the determination of T_g.

9.3.2.4 Step-wise Isothermal Frequency Scans

For a more comprehensive examination of mechanical deformational mechanisms, it is preferable to carry out frequency scans at a series of isothermal temperatures. In a single experiment, it is possible to cover four or more decades of frequency. This allows a highly detailed exploration of relaxation processes in the material under study. It can also be argued that the sample temperature is more precisely defined, as the frequency range is scanned after thermal equilibrium has been achieved. Figure 9.16 show plots of storage modulus, E' and $\tan\delta$ against frequency for an elastomeric material. Each curve represents a frequency scan made at the indicated isothermal temperature.

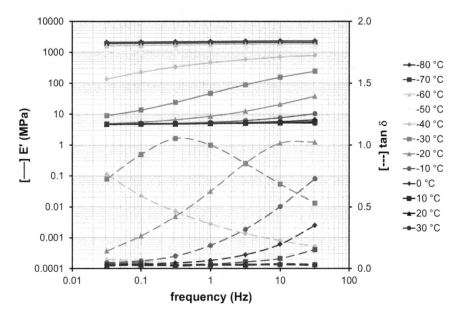

Figure 9.16 E' and $\tan\delta$ vs. frequency for an elastomer at different temperatures.

9.3.2.5   Time-temperature Superposition

As discussed earlier, the observed temperature at which the glass–rubber transition occurs depends upon the timescale over which one investigates molecular mobility (this applies to all methods of determining this parameter). To a first approximation, the process can be treated as a simple thermally activated effect and the relationship between the temperature of maximum mechanical damping (T_{peak}) and the timescale (or frequency, f) of the applied forcing variable (in this case mechanical deformation) can be analysed using a simple Arrhenius expression:

$$\ln(f) = \ln(A) - E_a/(R\ T_{peak}) \qquad (9.16)$$

where E_a is the apparent activation energy for the process and R is the gas constant.

A more rigorous approach recognises that the glass–rubber transition is a co-operative effect and does not conform to the simple model described. A common method for treating such a response is to superimpose data collected at different temperatures and frequencies onto one smooth curve. With reference to Figure 9.16, it can be seen that if a curve at one temperature is chosen as a reference point then data from other temperatures can be shifted in frequency to produce a smooth continuous change in properties spanning a wide frequency range. Ideally, both the moduli and damping factor data should produce good overlays (Figure 9.17). The relationship between the frequency shift (a) at a specific temperature (T) and the reference temperature (T_{ref}) is usually expressed in terms of the Williams–Landel–Ferry (WLF) equation:[11]

$$\log[a_T] = C_1(T - T_{ref})/(C_2 + T - T_{ref}) \qquad (9.17)$$

where C_1 and C_2 are constants.

Time-temperature superposition is a means of extending the frequency range of dynamic mechanical data and has applications for the evaluation of materials for acoustic damping properties. Note the shift is only strictly valid above T_g. This is due to the physical ageing of glasses that occurs below T_g.

Time-temperature superposition should only be used on single glass transition relaxation processes. Where there is an overlapping β process, for example, this must be mathematically removed from the experimental data before data analysis is performed. If there is doubt over whether a single relaxation exists, a Wicket plot can be

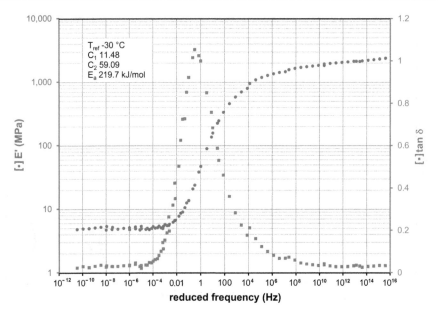

Figure 9.17 "Master curve" of E' and tan δ for data from Figure 9.16 reduced to $-30\,°C$ (inset shows fitting parameters for eqn (9.17)).

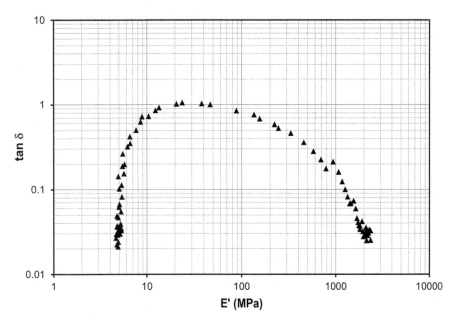

Figure 9.18 Wicket plot for data in Figure 9.16.

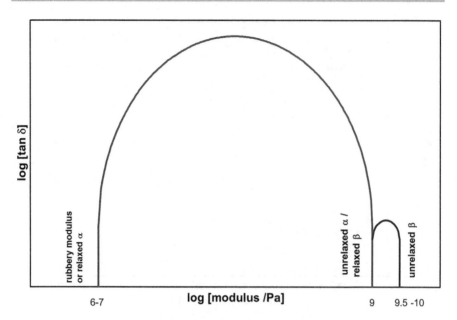

Figure 9.19   Theoretical Wicket plot.

constructed. The data shown in Figure 9.16 have been used to generate a Wicket plot (Figure 9.18), which exhibits a clean single "hoop" shape. This confirms the presence of a single relaxation. When more than one relaxation is present, then the curve is significantly distorted. A theoretical Wicket plot is shown in Figure 9.19. The values on the storage modulus axis (E') are shown as the equilibrium rubbery modulus value (usually 10^6–10^7 Pa) and then moving to the right, the unrelaxed α/relaxed β modulus may be observed followed by the unrelaxed β. The storage modulus values above 10^9 Pa are typical for a glassy material. This clearly demonstrates why overlapping relaxations cannot be treated by a WLF analysis. More elaborate models exist to describe the glass transition process, but they typically only have greater precision when the measured frequency range covers six decades or more.[12,13]

9.3.3   Dielectric Techniques

Measurements of electrical properties are particularly sensitive probes of the mobility of ions and dipoles within a specimen. Even non-polar materials such as polyethylene often contain polar impurities, which give sufficient response for the behaviour of the specimen to be analysed by these methods. The experiments described next illustrate the specialised niche occupied by TSDC and DETA for obtaining insights into molecular mobility and energetics.

9.3.3.1 Thermally Stimulated Current Measurements

Polypropylene glycol (PPG) is a common ingredient in pharmaceutical formulations and the molecular dynamics of the molecule are of interest for drug delivery applications. The knowledge of the glass transition (T_g) behaviour is essential to predict the shelf life of frozen products since it is likely that the material will change over time and the stability of any active ingredient in this matrix cannot be guaranteed. Although DSC can be used to study the glass transition *via* the change in heat capacity that accompanies devitrification, this effect can often be rather small (particularly for materials with low amorphous content) and there may not be enough sensitivity to detect the glass transition without ambiguity. A feature of the thermally stimulated current technique is that the current flow is directly proportional to the strength of the electric field. Thus it is possible to "magnify" weak transitions by increasing the polarisation voltage. Figure 9.20 shows the results of a series of experiments where

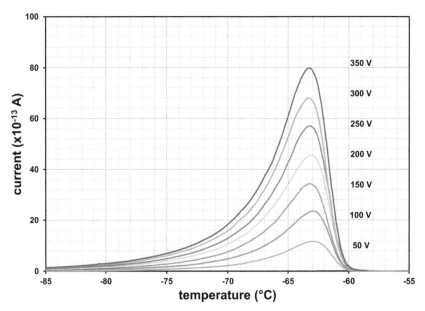

Figure 9.20   Global TSDC curve for polypropylene glycol polarised under different applied electric field strengths.

polypropylene glycol was cooled from −50 °C to −90 °C under the influence of an increasing applied field. Once the sample had been polarised, the electric field switched off and the depolarization current measured, it was warmed back up in the absence of any external field. The peaks observed indicate the T_g of the polypropylene glycol. These discharge curves are known as "global" depolarization curves—the more advanced technique of carrying out the polarisation step over a narrow temperature window in order to examine the time and temperature dependence of the discharge process will be examined later.

9.3.3.2 *Dielectric Thermal Analysis*

Thin films lend themselves to analysis by dielectric techniques and can often be tested with minimal sample preparation. Paints and other coatings can also be examined in this way, particularly if they are laid down on a metallic substrate that can then be used as one of the electrodes. Figure 9.21 shows plots of ε' and ε'' obtained at various frequencies for a sample of PET film. The glass–rubber transition of the polymer can be observed as a step increase in ε' and a peak in ε'' around 150 °C which shifts to higher temperatures as the measurement frequency increases.

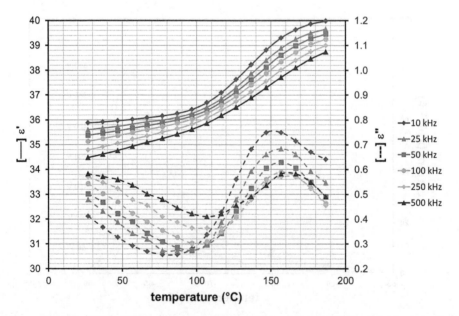

Figure 9.21 Dielectric analysis of PET film at different electrical field frequencies shown. Heating rate: 2 °C min^{-1}.

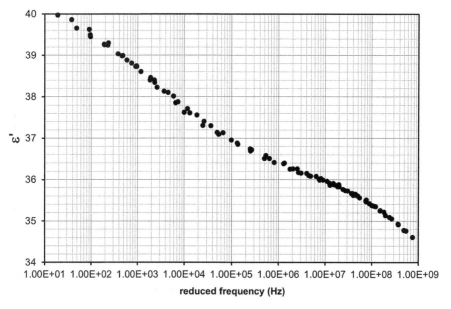

Figure 9.22 "Master curve" of data from Figure 9.21 with a reference temperature of 100 °C.

This is considerably higher than the T_g observed by DSC (typically around 70–80 °C) owing to the shorter timescale over which the dielectric technique probes molecular mobility. Also seen is the β transition of the polymer around 30 °C which is due to local movement of dipoles associated with the ester linkages. This also increases in temperature with increasing electric field frequency. In an analogous way to dynamic mechanical data, the dielectric results can be overlaid onto one smooth curve by using the principles of time-temperature superposition. Figure 9.22 shows such a curve using a reference temperature of 100 °C. Under these conditions, the dielectric properties of the material can be extrapolated into the radio and microwave frequency domain.

The shift factors used to generate such a master curve are plotted in Figure 9.23 against the reciprocal of the absolute temperature. This illustrates that the lower temperature β transition of PET obeys Arrhenius behaviour, with an apparent E_a of 650 kJ mol^{-1} whereas the glass–rubber (α) transition shows a WLF type dependence by its deviation from linearity, although a nominal E_a of 2500 kJ mol^{-1} can be assigned to this process. Studies of this type of behaviour can be used to distinguish between different types of relaxation phenomena in materials.

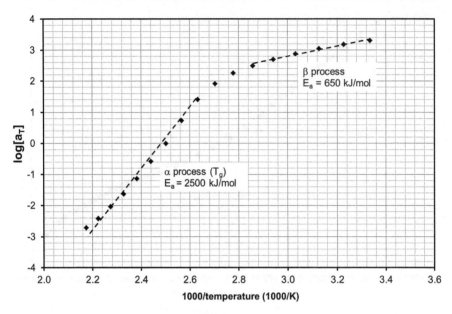

Figure 9.23   Plot of shift factors used to generate the master curve shown in Figure 9.22. The approximate "activation energies" for the α and β processes are indicated.

9.4   Applications

The following examples are used to demonstrate the broad range of applications of thermomechanical and thermoelectrical measurements.

9.4.1   Thermomechanical Analysis

Thermomechanical measurements are routinely used to investigate dimensional stability and this is particularly important in structures that will be exposed to a wide variation in temperature as part of their operational conditions. An example of this is shown in Figure 9.24 for an electrical power transformer. The device is encapsulated in a highly filled epoxy resin and specimens had exhibited cracking during use. Specimens of the epoxy were obtained by carefully sectioning the transformer and subjected to thermomechanical analysis. A heat/cool temperature programme was used whereby the specimen's change in length was monitored while it was cycled between ambient temperature and 125 °C at a rate of 5 °C min^{-1} (Figure 9.25). The heating curve shows an "S"-shaped change in dimensions that can be attributed to the epoxy going through its glass–rubber transition

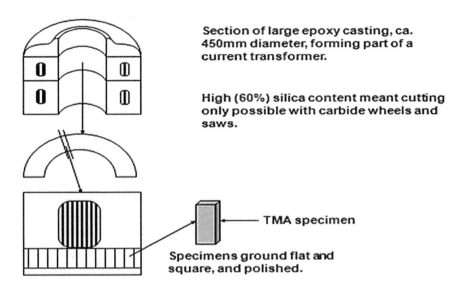

Section of large epoxy casting, ca. 450mm diameter, forming part of a current transformer.

High (60%) silica content meant cutting only possible with carbide wheels and saws.

TMA specimen

Specimens ground flat and square, and polished.

Figure 9.24 Diagram showing geometry of resin-encapsulated power transformer and location of samples taken for thermomechanical measurements.

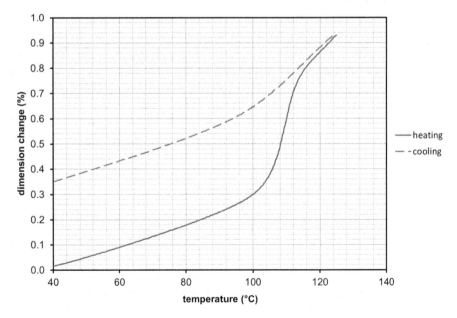

Figure 9.25 Typical TMA curve of sample showing heating and cooling response of epoxy resin.

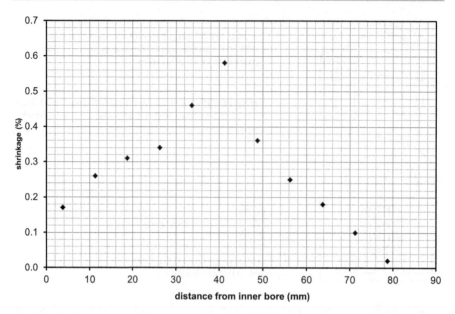

Figure 9.26 Graph showing shrinkage of epoxy resin samples derived from TMA curves as a function of distance from the inner bore of the transformer shown in Figure 9.25.

accompanied by some volumetric relaxation. The cooling curve shows the reverse process of vitrification with a residual difference in length on returning to the starting temperature. The T_g can be assigned for the cooling curve by extrapolating the linear sections of the length *vs.* temperature plot to their intersection around 100 °C.[23] The T_g on heating is less easy to define due to the relaxation effects. Nevertheless, by testing several samples taken across the section through the device, a map of residual stresses can be built up (Figure 9.26) and used to predict the location and likelihood of any mechanical failure.

9.4.2 Dynamic Mechanical Analysis

9.4.2.1 The Glass Transition of Reinforced or Filled Materials

With reference to Figure 9.15 and Table 9.2, it can be seen how a number of parameters can be used to define a value for T_g. These parameters may change unequally for different specimens of notionally the same material. Fibre reinforced composites are an excellent example of this phenomenon. Figure 9.27 shows the DMA curves of a composite and a sample of the base resin without fibre reinforcement and cured to the same degree. The tan δ peak is seen to decrease by around 7 °C for the composite sample and there is a lower

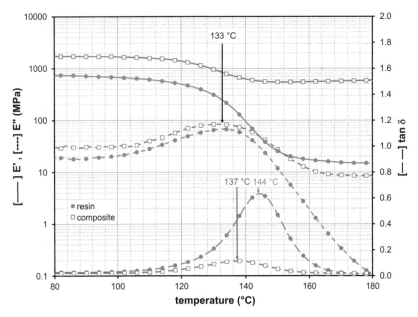

Figure 9.27 DMA curves of composite and neat resin.

peak magnitude, which is due to the fibre reinforcement. If the logarithmic onset of the change in E' is measured, this decreases by approximately 10 °C compared with the neat resin sample. These are consequences of geometry and do not reflect a different T_g in real terms. The contribution to E'' arises predominantly from the resin component, whereas E' will be increased significantly by the fibre content. Tan δ is calculated from the ratio E''/E' and is smaller for the composite. As a mathematical consequence, its peak position shifts to a lower temperature as well. The shape of the E'' curve is little affected by the fibre reinforcement, as would be expected, since this has a predominantly elastic effect. Its magnitude is slightly higher, due to the higher overall stiffness of the composite sample, but within ex-perimental error, the peak position is unchanged. This illustrates how carefully dynamic mechanical parameters should be specified, since one value shown here (tan δ) may condemn a sample as inferior and another (E'') would suggest it was within specification.

9.4.2.2 Glass Transition Measurement in Miscibility Studies

Polymer–polymer miscibility has great importance in many com-mercial systems and the glass transition is an excellent means to study this behaviour. When a material made from two or more components forms a continuous solid solution, it typically exhibits

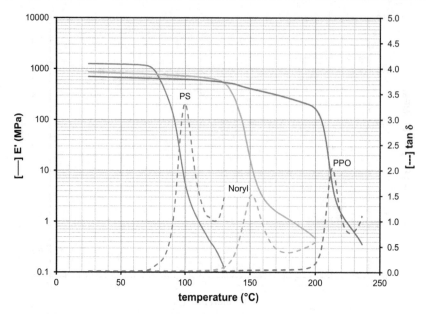

Figure 9.28   DMA curves of PS, PPO and Noryl™ (a 50 : 50 blend of the two polymers).

only one T_g.[14] If there is limited or no miscibility, then the original glass transition processes are observed. A broadening of the glass transition may be seen in one or more components where there is limited miscibility. Miscible behaviour is seen in Figure 9.28, which shows the DMA curves for polystyrene (PS), poly(2,6-dimethyl-1,4-phenylene ether) (commonly called "poly(phenylene oxide)" or PPO) and a commercial material Noryl™, which consists of a 50 : 50 blend of PS and PPO. These polymers are unusual in that they form a continuous solid solution across the entire composition range. Therefore, when the T_g of the blend is measured, it lies between that of the components, confirming the existence of a continuous solid solution. The value of T_g can be used to estimate the composition. The high sensitivity of DMA to the glass transition process makes this technique invaluable in the study of miscibility, especially for systems with a great excess of one component over the other where the T_g of the minority component would be difficult to detect by DSC.

9.4.2.3   DMA Measurements Under Controlled Relative Humidity

Many commercial instruments now offer accessories to provide controlled temperature and relative humidity environments. Figure 9.29 shows an example of two scanning humidity experiments on

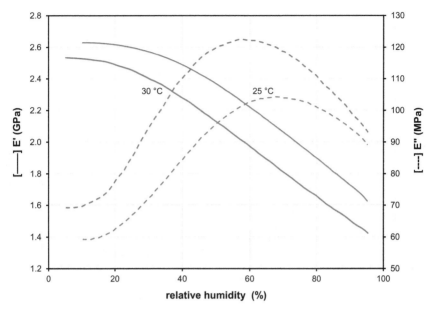

Figure 9.29 DMA data showing the effect of relative humidity on nylon 66 at indicated temperatures showing E' and E''.

nylon-6,6 film. One experiment was performed at 25 °C and the other at the slightly higher temperature of 30 °C. The plots show storage and loss moduli for these two temperatures as a function of relative humidity. The relative humidity was increased at 0.5% RH min^{-1} to ensure equilibrium in the film sample. The moduli can be seen to be decreasing as a function of increasing relative humidity and the loss modulus exhibits a peak consistent with a glass–rubber transition. The values of storage and loss modulus are lower at 30 °C than 25 °C, as would be expected, and the amount of water needed to reduce the T_g to 30 °C is less than that required at 25 °C.

The use of such environmental testing is extremely useful for many applications where materials experience the effects of moisture, *e.g.* paint films,[15] adhesives,[16] wound dressings[17] and polyelectrolyte membrane films (such as Nafion® as used in fuel cells[18]). There is considerable interest in studying the response of artists' materials to moisture for conservation purposes[19–21] and the DMA technique is not limited to the testing of samples in humid atmospheres: other vapours may be employed, or the samples may be completely immersed in a liquid.[22]

9.4.2.4 Isothermal Cure of Thermosetting Composite

DMA is regularly used to study the chemical reactions that lead to cross-linking of thermosetting resins, such as those used in the

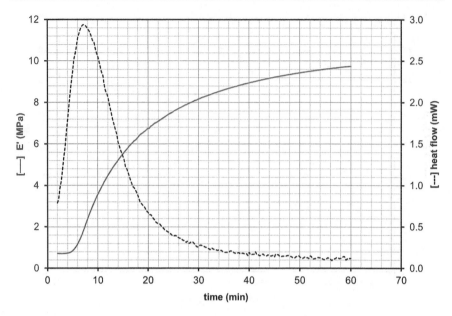

Figure 9.30   DMA and DSC curves of uncured pre-preg at 120 °C.

manufacture of composites. At high degrees of network formation, the rubbery cross-linked gel will vitrify into hard glassy material and the storage modulus will increase by several orders of magnitude. Plotting gel point and vitrification point against temperature and time leads to a time-temperature-transformation (or Gillham–Enns) diagram, which can be used to map out the curing of thermosetting polymers.[24]

Figure 9.30 shows an overlay of the DMA and DSC curves of a carbon fibre/epoxy resin composite as it cures isothermally at 120 °C. It can be seen that the peak in heat flow (corresponding to the maximum rate of reaction) coincides with the steepest change in stiffness but that the modulus of the sample continues to increase after 60 minutes even though the heat flow has subsided. DMA is more sensitive to the final stages of network formation during thermoset cure and is often used as a complementary technique to DSC.

9.4.3   Dielectric Techniques

9.4.3.1   *Thermally Stimulated Depolarisation Current Analysis (TSDCA)*

The global polarisation curves of polypropylene glycol are shown in Figure 9.20. In Figure 9.31, the thermal windowing technique has been employed to try to isolate individual relaxation processes that

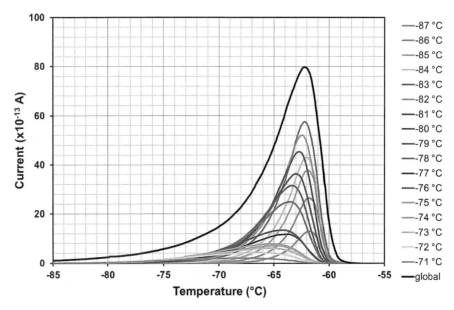

Figure 9.31 Thermal windowing experiment on polypropylene glycol.

are occurring. This results in a family of peaks arising from the discrete temperature ranges over which the sample was polarised. By analysing the thermal windowing peaks, it is possible to derive a series of distribution of relaxation times or "Buchi" lines (Figure 9.32) for each peak, which yields the E_a for the relaxation process in the polarisation window. These lines often appear to converge on a unique point—the so-called "compensation effect". However, the physical significance of the compensation point is surrounded by much controversy and is beyond the scope of this chapter.

9.4.3.2 Dielectric Thermal Analysis

Dielectric thermal analysis involves monitoring the viscosity of a system *via* its ability to store or transport electrical charge. Changes in the degree of alignment of dipoles and the ion mobility provide information pertaining to physical transitions in the material and to material properties such as viscosity, rigidity, reaction rate and cure state. By use of remote dielectric sensors, the measurements can be made in actual processing environments such as presses, autoclaves, and ovens. Dielectric measurements readily lend themselves to being carried out simultaneously with dynamic mechanical analysis when experiments are performed in compression or torsion. The sample is

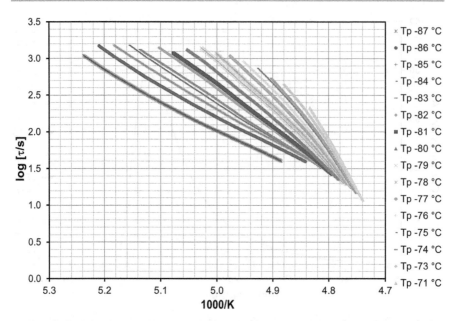

Figure 9.32 Compensation plot derived by analysing the relaxation curves in Figure 9.31.

usually mounted between parallel plates, which are used to apply the mechanical stress—electrical connections can be established to these and used to make a dielectric measuring cell.[25]

9.4.3.2.1 Water Mobility in Porous Polymers

Water is a ubiquitous substance, present in the environment and a compound that has an important effect on the mechanical and dielectric properties of many materials depending on the way that it is dispersed. The effect of humidity on the mechanical properties of materials has been discussed in Section 9.4.2.3. The water molecule has a particularly high dipole moment and thus is highly amenable to probing by dielectric techniques. Figure 9.33 shows a three-dimensional surface produced by plotting the dielectric loss tangent (tan δ) as a function of frequency and temperature for a porous polymer containing water. Two types of water can be identified: that contained within pores (which melt at $-10\,^\circ$C due to their small size), and more tightly bound water (which becomes mobile at lower temperatures) associated with polar groups on the polymer chain.[26]

9.4.3.2.2 Cavity Perturbation Methods

The dielectric properties of materials at radio and microwave frequencies are important in a number of applications ranging from the

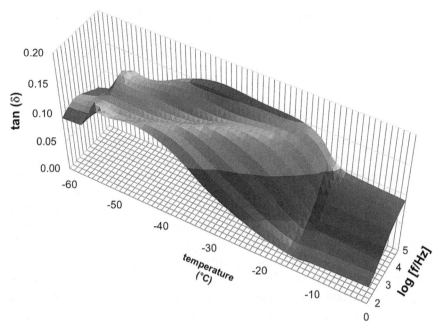

Figure 9.33 Three-dimensional surface plot of the dielectric loss tangent of a polymer film containing water in pores within its structure.

preparation of convenience foods to radar avoidance "stealth" technology. Whilst it is possible to extrapolate low frequency dielectric data into this region by means of temperature-time superposition, as was discussed in Section 9.3.3.2, it is preferable to measure the required properties directly using transmission line or cavity perturbation techniques. Such measurements involve measuring the frequency and power response of a tuned cylindrical cavity with and without the sample in place to derive the relative permittivity and dielectric loss factor. Furthermore, high power radio and microwave energy cannot only be used as a probe of dielectric response but also as a means of heating the specimen.[27] Such an approach can be used to investigate anomalous behaviour when materials are exposed to microwaves.[28]

9.5 Sample Controlled and Modulated Temperature Techniques

With the exception of the thermal windowing technique applied to TDSCA, we have so far only considered temperature profiles that are relatively simple in structure—either a linear rise and/or fall in temperature or a series of isothermal steps. In this section, we will briefly consider the benefits of more sophisticated strategies for the alteration of specimen temperature.

9.5.1 Rate Controlled Sintering of Ceramics

The compaction and sintering of high temperature refractory materials may be studied by thermomechanical analysis. In many cases, it is desirable that the ceramic changes in dimensions in a uniform manner. In order to achieve this, the rate of heating can be controlled by the rate of change of dimensions of the specimen. This can be done by heating the sample at a fixed heating rate and then stopping heating when the rate of change of length exceeds a certain limit. The process is allowed to continue isothermally until the rate falls below the limit and then heating is recommenced.[29] An alternative approach is illustrated in Figure 9.34 whereby a sample of a precursor for alumina is heated at a fixed rate of temperature rise ($10\,^\circ$C min^{-1}) until the rate of shrinkage reaches a pre-set value (-0.02 mm min^{-1}), thereafter the rate of temperature change is controlled so as to maintain this constant rate of shrinkage. The length change and

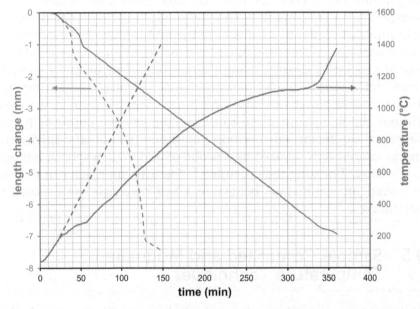

Figure 9.34 Rate-controlled sintering of alumina (solid line = constant rate of change of length, broken line = constant heating rate).

temperature profiles of a constant heating rate experiment are shown alongside those from the rate-controlled sintering experiment under the same conditions. It can be seen that although the constant heating rate experiment is quicker to perform, the rate-controlled experiment results in a smaller overall change in dimensions and finer crystal structure.[30] In a manner analogous to multiple linear heating rate kinetic methods used for thermogravimetry, kinetic models of sintering can be developed to permit simulations to be developed, these predictions can then be employed to develop heating regimes for furnaces used to fire ceramics.[31]

A similar technique has been described for DMA whereby the temperature program was controlled by constraining the rate of change of mechanical properties (*e.g.* storage modulus) to within certain limits. This approach was shown to be effective in resolving the multiple glass transitions of a polymer blend.[32]

9.5.2 Modulated Temperature Programs

In the same way that applying a temperature modulation to DSC can be used to separate thermally reversing thermal events (such as the glass transition) from thermally non-reversing ones (*e.g.* crystallisation and curing), the same principles can be applied to TMA. This approach enables one to separate reversible dimensional changes due to thermal expansion from irreversible effects such as creep or stress relaxation.[33–35]

Figure 9.35 shows such a modulated temperature TMA experiment. The specimen comprises the filled epoxy resin material that was discussed in Section 9.4.1. Instead of a linear heat–cool temperature cycle, the specimen is exposed to a sinusoidal rising temperature programme with an amplitude of 3 °C and period of 300 s superimposed on an underlying heating rate of 0.3 °C min^{-1}. The total rate of change in length is obtained by averaging the signals to remove the effect of the temperature modulation and differentiating the change in length with respect to temperature. The sharp peak is due to the volumetric relaxation of stresses within the material also seen in the heating part of the conventional experiment shown in Figure 9.25. A much cleaner change in the specimen's properties can be elucidated by dividing the amplitude of the length change by the amplitude of the temperature modulation. This gives the reversing rate of change of length also shown in Figure 9.35 (*c.f.* cooling curve in Figure 9.25) and allows the glass transition to be clearly defined by the step change in this parameter analogous to modulated temperature DSC.[36]

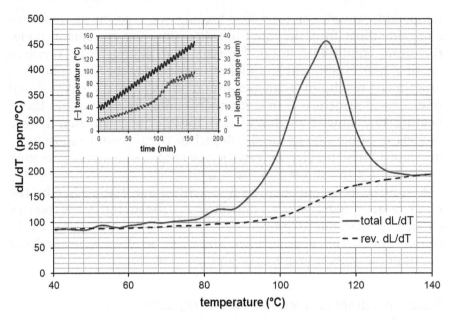

Figure 9.35　Modulated temperature TMA of epoxy resin sample taken as indicated in Figure 9.24 (inset shows raw temperature and length *vs.* time data).

Modulated temperature DMA has been developed as a means of investigating the reversible melting of polymers.[37,38] A sinusoidal heating program has also been employed in TSC to separate reversible pyroelectric currents from non-reversible thermally stimulated discharge of heated dielectric materials.[39]

9.6　Localised Thermomechanical and Dynamic Mechanical Measurements

One of the drawbacks of all of the thermomechanical and dynamic mechanical techniques discussed is that they require relatively large specimens for testing. Furthermore, the results of such measurements cannot distinguish between a small change in the bulk properties of the material and a large change in the behaviour of a minor component in the specimen under examination. In order to overcome this drawback, a family of techniques, loosely termed "micro-thermal analysis", has been developed, which combines several forms of thermal and chemical characterisation with an imaging technique based upon high resolution profilometry to afford a means of analytical microscopy with high spatial resolution.[40]

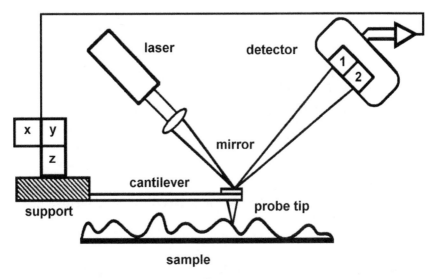

Figure 9.36 A schematic representation of an atomic force microscope.

The basis of micro-thermal analysis is an atomic force microscope (AFM), as illustrated in Figure 9.36. Piezo-electric elements are used to move a sharp tip on the end of a cantilever across the surface of the specimen while an optical lever formed by a laser spot reflected from the back of the cantilever provides a force-feedback loop so that the tip follows the topography of the sample's surface without causing any damage. The spatial resolution of the image is limited to the tip radius and, with care, can approach the atomic level.

Further refinements to the imaging technique can monitor the lateral twisting of the cantilever whilst it is rastered over the sample. This gives an indication of the frictional interaction between the tip and the surface.[41] Intermittent contact modes can be used to produce images based upon the force needed to lift the tip away from the surface of the sample. The pair of images shown in Figure 9.37 illustrate the changes in contrast brought about by scanning a sample of a polystyrene/poly(methyl methacrylate) blend on a heated stage at 50 °C and 110 °C. There is a dramatic increase in tip–sample adhesion for the dispersed polystyrene phase above its glass transition temperature (100 °C) whereas the poly(methyl methacrylate) matrix exhibits no such effect.[42]

Rather than heat the whole specimen, it is possible to employ a special probe whereby the tip can be heated. Three such designs are shown in Figure 9.38 (with the tip uppermost) so as to illustrate the working principle of each configuration. At the base of each cantilever

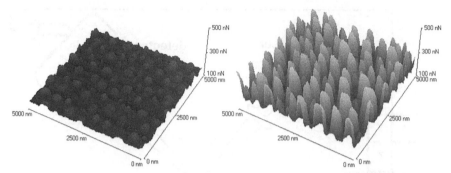

Figure 9.37 Pulsed force mode (pull-off force) images of a polystyrene/
poly(methyl methacrylate) blend at 50 °C (left) and 110 °C
(right).

Figure 9.38 Schematic diagrams of resistive scanning thermal microscopy
probes: (a) Wollaston wire type (b) micro-machined coated Si
cantilever and (c) doped silicon probe.

is a semi-circular carrier used to attach it to the scanner. The probe
shown in Figure 9.38a is fashioned from Wollaston process wire,
which consists of a 75 μm diameter silver wire surrounding a 5 μm
diameter core of platinum/10% rhodium alloy. The wire is bent to
form a sharp loop and secured into shape with a bead of epoxy resin.
The silver layer is then etched away at the apex to reveal the platinum
filament that forms the major electrical resistance element (approximately 2 Ω) in the assembly and acts as a temperature sensor and
heater. A reflective mirror is glued on the wires to serve as a target for
the laser.[43] The probe shown in Figure 9.38b employs a thin resistance element deposited across the apex of a silicon nitride pyramid,
similar to a conventional atomic force microscope cantilever,[44]
whereas the probe shown in Figure 9.38c combines elements from
the first and second embodiments so that a small pyramid of material
is heated by current passing down each arm of the cantilever.[45]

   All three probes can be operated in two modes: (a) as a passive
thermo-sensing element (by measuring its temperature using a small

excitation current)[46] or (b) as an active heat flux meter. In the latter case, a larger current (sufficient to raise the temperature of the probe above that of the surface) is passed through the heater. The power required to maintain a constant temperature gradient between the tip and sample is monitored by means of an electrical bridge circuit. In essence, this is equivalent to a power compensation calorimeter and allows an image to be constructed based upon the thermal conductivity of the sample's surface.

Figure 9.39 shows the thermal conductivity contrast image of a blend of two polymers—poly(vinyl acetate) and poly(vinyl butyral). There is a clear discrimination between the matrix and the dispersed phase. The results of a measurement on the bulk material by modulated-temperature DSC are shown in Figure 9.40. The derivative of the reversing heat capacity (dC_p/dT) shows a peak with a shoulder arising from the closely spaced glass–rubber transitions of the two polymers. The data can be fitted to two Gaussian peaks (as shown) in order to quantify the amount of each polymer present and investigate any interfacial effects.[47] Thus both microscopy and

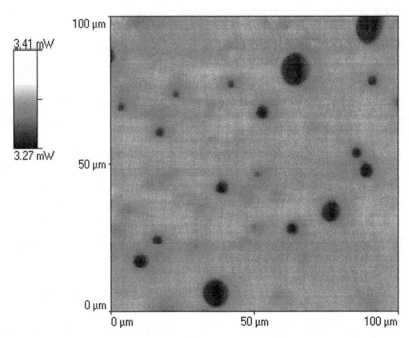

Figure 9.39 Thermal conductivity contrast image of a blend of poly(vinyl acetate) and poly(vinyl butyral). The dark areas represent regions of low thermal conductivity.

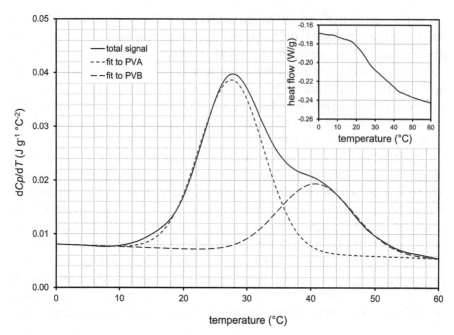

Figure 9.40   First derivative of heat capacity with respect to temperature for the two-component polymer blend shown in
Figure 9.39 (solid line—raw data, dashed lines—fitted data
to individual components). The inset shows the conventional DSC curve.

calorimetry confirm the existence of two components but there
remains some ambiguity over the identity of the matrix and
dispersed phase.

By combining the accurate positioning of the piezoelectric scanner,
the force feedback loop of the optical lever and the ability to heat the
tip of the probe, it is possible to perform localised thermomechanical
analysis of selected areas of a sample. Four such measurements on
the sample shown in Figure 9.39 are shown in Figure 9.41. These
confirm that the matrix is poly(vinyl acetate) and the dispersed phase
is poly(vinyl butyral), from the softening temperatures recorded by the
probe displacement.

Heating rates of the order of $10–100\,^{\circ}\mathrm{C}\ \mathrm{s}^{-1}$ or higher can be used
because of the low thermal mass of the probe and small volume of
sample that is heated ($<500\ \mu\mathrm{m}^{3}$). Furthermore, it is possible to carry
out evolved gas analysis of any volatiles released during heating[48] or
exploit the ability of the probe to respond to temperature changes by
irradiating the surface with infrared light in order to record infrared

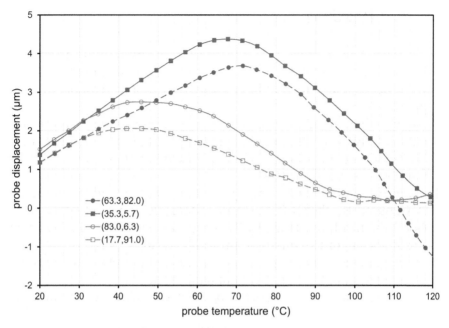

Figure 9.41 Localised thermomechanical analysis of regions of the sample shown in Figure 9.39. Open symbols represent measurements on the matrix and filled symbols represent those on the dispersed phase.

spectra at an ultra-high spatial resolution.[49] Micro-thermal analysis has the ability to map thermal properties in two or even three dimensions,[50] with numerous applications to many areas of materials characterisation, and the reader is encouraged to refer to more specialist publications for further details.

9.7 Summary

The description of thermomechanical and thermoelectrical measurements in a concise chapter such as this is an ambitious exercise. We have attempted to introduce a wide range of methods and applications with the intention of illustrating the diversity of this field whilst emphasising the connections between the inter-related techniques. With the exception of TMA (often belittled due to its simplicity), these methods are often promoted as some of the more "advanced" thermal analysis techniques. It is hoped that the preceding pages will dispel such preconceptions.

Further Reading

H. E. Bair, A. E. Akinay, J. D. Menczel, R. B. Prime and M. Jaffe, in *Thermal Analysis of Polymers, Fundamentals and Applications*, ed. J. D. Menczel and R. B. Prime, John Wiley & Sons, Hoboken, 2009, pp. 319–386.

S. A. B. Allen, in *Handbook of Thermal Analysis and Calorimetry*, ed. M. E. Brown, Elsevier, Amsterdam, 1998, vol. 1, pp. 401–422.

D. Q. M. Craig, *Dielectric Analysis of Pharmaceutical Systems*, Taylor and Francis, London, 1995.

J. Duncan, in *Principles and Applications of Thermal Analysis*, ed. P. Gabbott, Blackwell Publishing, Oxford, 2008, pp. 119–163.

J. Goodwin and R. W. Hughes, *Rheology for Chemists: An Introduction*, Royal Society of Chemistry, Cambridge, 2000.

J. P. Ibar, *Fundamentals of thermal stimulated current and relaxation map analysis*, SLP Press, New Canaan, 1993.

J. D. Ferry, *Viscoelastic Properties of Polymers*, Wiley, New York, 3rd edn, 1980.

N. G. McCrum, B. E. Read and G. Williams, *Anelastic and Dielectric Effects in Polymeric Solids*, Dover, New York, 1991.

K. P. Menard, *Dynamic Mechanical Analysis: A Practical Introduction*, CRC Press, Boca Raton, 2nd edn, 2008.

L. E. Nielsen and R. F. Landel, *Mechanical Properties of Polymers and Composites*, Dekker, New York, 2nd edn, 1993.

A. T. Riga and C. M. Neag, *Materials Characterization by Thermo-mechanical Analysis*, ASTM STP 1136, American Society for Testing and Materials, Philadelphia, 1991.

J. P. Runt and J. J. Fitzgerald, *Dielectric Spectroscopy of Polymeric Materials: Fundamentals and Applications*, American Chemical Society, Washington DC, 1997.

M. Reading and P. J. Haines, in *Thermal Methods of Analysis: Principles, Applications and Problems*, ed. P. J. Haines, Blackie, Glasgow, 1995, pp. 123–160.

B. B. Sauer, in *Handbook of Thermal Analysis and Calorimetry*, ed. Z. D. Cheng, Elsevier, Amsterdam, 2002, vol. 3, pp. 653–711.

M. T. Shaw and W. J. MacKnight, *Introduction to Polymer Viscolelasticity*, John Wiley & Sons, Hoboken, New Jersey, 3rd edn, 2005.

R. E. Wetton, in *Handbook of Thermal Analysis and Calorimetry*, ed. M. E. Brown, Elsevier, Amsterdam, 1998, vol. 1, pp. 363–399.

Acknowledgements

The authors are grateful to Dr Milan Antonijevic and Dr Samuel Owusu-Ware (University of Greenwich) for providing TSC data.

References

1. ASTM E2347-11 *Standard Test Method for Indentation Softening Temperature by Thermomechanical Analysis*, ASTM International, West Conshohocken, 2011.
2. ASTM E2092-13 *Standard Test Method for Distortion Temperature in Three-Point Bending by Thermomechanical Analysis*, ASTM International, West Conshohocken, 2013.
3. ASTM E1363-13 *Standard Test Method for Temperature Calibration of Thermomechanical Analyzers*, ASTM International, West Conshohocken, 2013.
4. ASTM E2206-11 *Standard Test Method for Force Calibration of Thermomechanical Analyzers*, ASTM International, West Conshohocken, 2011.
5. ASTM E2113-13 *Standard Test Method for Length Change Calibration of Thermomechanical Analyzers*, ASTM International, West Conshohocken, 2011.
6. ASTM E831-14 *Standard Test Method for Linear Thermal Expansion of Solid Materials by Thermomechanical Analysis, ASTM International, West Conshohocken*, 2014.
7. D. M. R. Georget, A. Ng, A. C. Smith and K. W. Waldron, *J. Sci. Food Agric.*, 1998, **78**, 73.
8. A. Nikolaidis and T. P. Labuza, *J. Thermal Anal. Calorim.*, 1996, **47**, 1315.
9. D. M. R. Georget and A. C. Smith, *J. Thermal Anal. Calorim.*, 1996, **47**, 1377.
10. L. Woo, S. P. Westphal, S. Shang and M. Y. K. Ling, *Thermochim. Acta*, 1996, **284**, 57.
11. M. L. Williams, R. F. Landel and J. D. Ferry, *J. Am. Chem. Soc.*, 1955, 77, 3701.
12. G. Williams and D. C. Watts, *Trans. Faraday Soc.*, 1970, **66**, 80.
13. S. Havriliak and S. Nagami, *J. Polym. Sci. C*, 1966, **14**, 99.
14. W. J. MacKnight, F. E. Karasz and J. R. Fried, in *Polymer blends*, ed. D. R. Paul, S. Newman, Academic Press, 1978, vol. 1, pp. 186–243.

15. G. M. Foster, S. Ritchie and C. Lowe, *J. Therm. Anal. Calorim.*, 2003, **73**, 119.
16. G. LaPlante and P. Lee-Sullivan, *J. Appl. Polym. Sci.*, 2005, **95**, 1285.
17. J. F. Mano, *Macromol. Biosci.*, 2008, **8**, 69.
18. F. S. Bauer, S. Denneler and M. Willert-Porada, *J. Polym. Sci. Part B: Polym. Phys.*, 2005, **43**, 786.
19. N. S. Cohen, M. Odlyha and G. M. Foster, *Thermochim. Acta*, 2000, **365**, 111.
20. M. Odlyha, in *Handbook of Thermal Analysis and Calorimetry*, ed. M. E. Brown and P. K. Gallagher, Elsevier, 2003, vol. 2, pp. 47–92.
21. M. Odlyha, Q. Wang, G. M. Foster, J. de Groot, M. Horton and L. Bozec, *J. Therm. Anal. Calorim.*, 2005, **82**, 627.
22. D. M. Price, *Thermochim. Acta*, 1997, **294**, 127.
23. C. M. Earnest, in *Assignment of the Glass Transition*, ed. R. J. Seyler, ASTM STP 1249, American Society for Testing and Materials, Philadelphia, 1994, pp. 75–87.
24. J. K. Gillham and J. B. Enns, *Trends Polym. Sci.*, 1994, **2**, 406.
25. B. Twombly and D. D. Shepard, *Instrum. Sci. Technol.*, 1994, **22**, 259.
26. R. Pelster, *Phys. Rev. B: Condens. Matter Mater. Phys.*, 1999, **59**, 9214.
27. A. Nesbitt, P. Navabpour, C. Nightingale, T. Mann, G. Fernando and R. J. Day, *Meas. Sci. Technol.*, 2004, **15**, 2313.
28. J. G. P. Binner, G. Dimitrakis, D. M. Price, M. Reading and B. Vaidhyanathan, *J. Therm. Anal. Calorim.*, 2006, **84**, 409.
29. O. Toft Sørensen, *Thermochim. Acta*, 1981, **50**, 163.
30. H. Palmour III, in *Science of Sintering*, ed. D. P. Uskoković, H. Palmour III and R. M. Spriggs, Springer, USA, 1989, pp. 337–356.
31. J. Opfermann, J. Blumm and W.-D. Emmerich, *Thermochim. Acta*, 1998, **318**, 213.
32. M. Reading, in *Thermal Analysis–Techniques & Applications*, ed. E. L. Charsley and S. B. Warrington, The Royal Society of Chemistry, Cambridge, 1992, pp. 127–155.
33. D. M. Price, *Thermochim. Acta*, 2000, **357/358**, 23.
34. P. Kamasa, P. Myslinski and M. Pyda, *Thermochim. Acta*, 2006, **442**, 48.
35. R. A. Shanks, *J. Thermal Anal. Calorim.*, 2011, **106**, 151.
36. D. M. Price, in *Material Characterization by Dynamic and Modulated Thermal Analytical Techniques*, ed. A. T. Riga and

L. H. Judovits, ASTM STP 1402, American Society for Testing and Materials, West Conshohocken, PA, 2001, pp. 103–114.

37. A. Wurm, M. Merzlyakov and C. Schick, *Coll. Polym. Sci.*, 1998, **276**, 289.
38. A. Wurm, M. Merzlykov and C. Schick, *Thermochim. Acta*, 1999, **330**, 121.
39. E. J. Sharp and L. E. Garn, *J. Appl. Phys.*, 1982, **53**, 8980.
40. D. M. Price, M. Reading, A. Hammiche and H. M. Pollock, *Int. J. Pharm.*, 1999, **192**, 85.
41. M. E. McConney, S. Singamaneni and V. V. Tsukruk, *Polym. Rev.*, 2010, **50**, 235.
42. D. B. Grandy, D. J. Hourston, D. M. Price, M. Reading, G. Goulart Silva, M. Song and P. A. Sykes, *Macromolecules*, 2000, **33**, 9348.
43. R. J. Pylkki, P. J. Moyer and P. E. West, *Jpn. J. Appl. Phys.*, 1994, **33**, 784.
44. G. Mills, J. M. R. Weaver, G. Harris, W. Chen, J. Carrejo, L. Johnson and B. Rogers, *Ultramicroscopy*, 1999, **80**, 7.
45. B. W. Chui, T. D. Stowe, T. W. Kenny, H. J. Mamin, B. D. Terris and D. Rugar, *Appl. Phys. Lett.*, 1996, **69**, 2767.
46. A. Majumdar, *Microelectron. Reliab.*, 1998, **38**, 559.
47. M. Song, D. J. Hourston, H. M. Pollock, F. U. Schäfer and A. Hammiche, *Thermochim. Acta*, 1997, **304/305**, 335.
48. D. M. Price, M. Reading, R. M. Smith, A. Hammiche and H. M. Pollock, *J. Therm. Anal. Calorim.*, 2001, **64**, 309.
49. A. Hammiche, L. Bozec, M. Conroy, H. M. Pollock, G. Mills, J. M. R. Weaver, D. M. Price, M. Reading, D. J. Hourston and M. Song, *J. Vac. Sci. Technol., B: Microelectron. Nanometer Struct.-Process., Meas., Phenom.*, 2000, **18**, 1322.
50. L. Harding, J. Wood, M. Reading and D. Q. M. Craig, *Anal. Chem.*, 2007, **79**, 129.

10 Simultaneous Thermal Techniques

Ian J. Scowen*[a] and Richard Telford[b]

[a] School of Chemistry, University of Lincoln, Brayford Pool, Lincoln
LN6 7TS, UK; [b] School of Chemistry and Forensic Sciences, University of
Bradford, Bradford BD7 1DP, UK
*Email: iscowen@lincoln.ac.uk

10.1 Introduction and Principles

Thermal analysis (TA) is well established as a powerful suite of techniques for the analysis of materials as diverse as pharmaceuticals,[1,2] polymers,[3,4] minerals,[5,6] clays,[7,8] soils,[9,10] catalysts,[11,12] and adhesives.[13,14] It has even been extended to an exotic materials context (*e.g.* Mars mineralogy[15]) and into the kinetic and thermodynamic interrogation of systems.[16] However, single technique analyses inherently limit the interrogation of materials and application of additional analytical probes in the experiment offers alternative, and often complimentary, information. In particular, thermal analysis cannot directly provide access to molecular and/or morphological information and so the addition of techniques able to provide this is highly attractive.

Furthermore, deploying several probes to a single experiment, ensuring identical composition and thermal history of the sample, offers a clear advantage over a series of separate experiments with different samples. Thermal analytical techniques have been successfully combined with a diverse range of analytical probes, several of which have become available as 'hyphenated' techniques with

Principles of Thermal Analysis and Calorimetry: 2nd Edition
Edited by Simon Gaisford, Vicky Kett and Peter Haines
© The Royal Society of Chemistry 2016
Published by the Royal Society of Chemistry, www.rsc.org

manufacturers supplying fully integrated instrumental platforms. This chapter will consider the increasing diversity of analytical methods deployed alongside TA methods and evaluate the complementarity of information available in current instrumentation.

It is perhaps helpful to classify modes of combined measurement within thermal analysis. Two primary categories emerge from a review of the techniques available:

(1) Simultaneous measurement including analysis of the material *in situ* in the calorimeter. Primarily, simultaneous thermal methods have been applied but there is an increasing application of microscopic and spectroscopic methods, representing some of the major advances in recent years.

(2) Sequential analysis involving evolved gases from the sample as the thermal analysis experiment proceeds. This mode of measurement is closely related to the 'hyphenation' of instrumentation increasingly applied in analytical science, *e.g.* the array of detection modes that can be appended to chromatographic separations: GC-MS, LC-UV, LC-MS *etc.* Several analytical methods have been applied to analyse the evolved gas stream and, in recent times, mass spectrometry and vibrational spectroscopy have become important in these applications.

10.2 Simultaneous Thermal Analysis

Simultaneous thermal analysis (STA) most commonly refers to the combination of thermal analytical techniques, in which two or more measurements are made on the same sample, at the same time. The complementary advantage of the simultaneous measurement of caloric and mass change information has led to the pre-eminence of DSC-TGA and DTA-TGA as the most widely established STA techniques and several commercial instruments are available. The key requirement for STA in these contexts is to obtain a thermocouple output from the sample without affecting the action of the balance. Several configurations are possible including 'top loading' balances and beam balances (Figure 10.1). While, in principle, there is an inherent conflict for simultaneous measurement—caloric information requires relatively small samples and high heating rates, and mass change information is favoured by relatively large samples and slow heating rates—the development in sensitivity of modern balances to sub-microgram changes has largely overcome these issues.

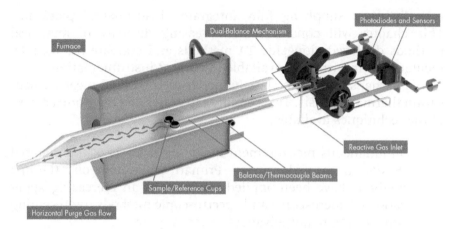

Figure 10.1 Schematic representation of a simultaneous TGA/DSC (copyright TA Instruments Waters L.L.C. Used with permission).

Simultaneous measurement, in particular the exact measurement
of weight loss during DSC analysis, opens opportunities to extend the
DSC and DTA experiments in quantitative analysis including deducing the energetics of complex decompositions,[17,18] measurements
of heats of vaporisation[19] and accounting for mass loss in the DSC
baseline.[20]

10.3   Evolved Gas Detection and Evolved Gas Analysis

The value of analysis of evolved gases produced during TA experiments has long been recognised and has informed a diverse array of
materials studies including examples in the analysis of organic materials, primarily pharmaceutical products and polymers[21,22] and inorganic materials including ceramics[23] and minerals.[15,24]

In the context of thermal experimentation, evolved gas detection
(EGD) and evolved gas analysis (EGA) are differentiated and are for-
mally defined by the International Union of Pure and Applied
Chemistry (IUPAC) in the *Compendium of Analytical Nomenclature*.[25]

In essence, EGD provides a means for detection of a gas and aids in
the confirmation of whether a TA signal can be associated with gas
loss or otherwise. The most commonly used detector systems are non-
specific and often correspond with detection systems found in simpler gas chromatographs *e.g.* thermal conductivity detectors, gas
density detectors, ionisation detectors *etc.* As a general principle,

detectors are placed as close to the sample as possible to reduce time lags, condensation and secondary reactions in the gas phase. Detector response times and sensitivity are key to informing appropriate gas flow requirements to ensure appropriate sampling of the experiment.

The term EGA is used when the gas analysis relates to qualitative and/or quantitative (molecular) speciation of the evolved gas stream. The use of EGA has generally superseded EGD and forms the basis for hyphenated TA instrumentation, most commonly with thermogravimetry (TG) as the sole thermal technique or in its combination with differential thermal methods (particularly DSC or DTA) in the hyphenation of simultaneous thermal instruments, *i.e.* STA-EGA. EGA has the capability to extend significantly the utility of the TG experiment, offering the potential for resolving multiple volatile components produced from a single weight loss step, and, with techniques such as mass spectrometry, substantially enhancing the sensitivity of the detector revealing processes occurring that may be unobservable by the thermal approach alone. In principle, EGA encompasses methods applied discontinuously, *i.e.* the off-line analysis of gases collected with adsorbent tubes,[26] cold traps[27] or bubbler solutions[28] from the TG experiment. In the modern context, however, EGA is associated with in-line (hyphenated) gas analysis methods and Fourier transform infrared spectroscopy (FTIR) and mass spectrometry are the major techniques deployed for molecular speciation of the evolved gases.

10.3.1 Thermal Analysis-mass Spectrometry

Mass spectrometry characterises gaseous ions by their mass/charge (m/z) ratio,[25] providing molecular identification from ions representing the molecular species (the molecular ion, alongside characteristic ions from its fragmentation) and quantification from the ion current at a particular m/z. Mass spectrometers (MS) comprise an ion source and an ion detector (mass analyser) operating under a high vacuum. The first reports of TG-MS couplings appeared in the mid-1960s,[29] including time of flight (TOF)[30] and quadrupole (QMS)[31] mass analysers. Coupling the instrumentation requires some consideration. Commonly, TA is performed under the flow of a purge gas, providing the means to 'sweep' the evolved gases to the mass spectrometer. However, this flow has to be transferred and effectively sampled to achieve the pressure drop required for effective operation of the mass spectrometer—typically, a reduction of pressure from atmospheric (*ca.* 10^3 mbar) in the thermobalance to high vacuum

$(10^{-5}$ mbar) in the MS. Transfer of the gas between the TG thermo-balance and the mass spectrometer is generally achieved with an appropriately short, heated transfer line constructed of a suitably inert material (usually silica). In this way, losses in resolution of thermal events arising from transfer times and/or diffusion effects in the gas flow, or, changes in molecular composition arising from condensation and/or reaction of the gaseous species are minimised.

Addressing the substantial pressure change required is less trivial and several designs for suitable interfaces have been investigated including metering valves,[32] capillaries,[33,34] orifice designs,[35,36] jet separators[37] and molecular leak interfaces.[38,39] The majority of commercial instrumentation is based on capillaries with a valve and/ or a pinhole diaphragm,[40] whereby a narrow heated capillary is presented to the evolved gases, either directly adjacent to the sample or to the exhaust of the instrument, drawing in a reduced flow (1–2 mL min^{-1}) to maintain the vacuum in the MS and allow the re-maining purge flow from the TA instrument to bypass to waste. The mass spectrometers in commercial systems for TG-MS most com-monly feature electron impact (EI) ionisation although chemical ionisation (CI) is also available. Mass analysers with TOF and QMS designs feature[32–39,41,42] although QMS is by far the most widely de-ployed. QMS instruments have scanning rates comfortably compat-ible with the speed of transitions observed in TA and benefit from relatively small footprints and cheap production costs. They do, however, have lower mass resolution leading to an inability to resolve gases of the same nominal mass (*e.g.* N_2 and CO at *m/z* 28) and require careful choice of purge gases.[40] TOF mass spectrometers are generally used only in specific applications exploiting their higher resolution and scanning speeds. Magnetic sector instruments appear to be abandoned completely in routine laboratory contexts. TA-MS instru-mentation (including interfacing) is currently provided by the ma-jority of manufacturers of thermal instrumentation, often with third party mass spectrometers.

There has been a steady progression in the number of published research papers reporting the development and/or use of EGA using MS. The early developments were reviewed by Dollimore *et al.*[43] and, more recently, by Raemaekers and Bart.[40] Modern applications are numerous and cover several application areas: polymers,[44,45] miner-alogy,[15] clays,[46] catalysis,[47] zeolites,[48] environmental chemistry,[49] pharmaceutical analysis[50] and petrochemicals.[51] TG-MS has particu-lar utility in identifying residual solvents and in characterising solvate (including hydrate) forms of materials. Quantitative application of

TG-MS is an increasingly important utility in monitoring of hetero-geneous catalyst efficiency,[47,48] in detailed elucidation of reaction stoichiometry[44,49] and decomposition.[52]

The thermal decomposition of calcium oxalate dihydrate can be used as an example to illustrate the principles of TG-MS analysis. Mass losses in the TG curve for this material correspond with changes in ion currents for MS channels at m/z 18, 28 and 44 (Figure 10.2). These confirm the identity of the molecular product at each step of the decomposition: loss of water (m/z 18) in the dehydration of the solid is followed by decomposition of the oxalate to carbonate (loss of CO, m/z 28), followed by decomposition of carbonate to the oxide (loss of CO_2, m/z 44). It is also apparent that a small amount of oxidation of CO to CO_2 is occurring, invoking a signal at m/z 44 concurrently with the evolution of CO. The presence of m/z 28 during the final step of the decomposition is due to fragmentation of CO_2 in the MS.

Accounts of MS hyphenated to simultaneous thermal techniques (STA) are also emerging. Grishchenko *et al.*[53] use TGA-DSC-MS as part of their characterisation of a synthetically produced inorganic com-pound close to a Friedel's Salt ($[Ca_2Al(OH)_6]Cl_{0.90}(CO_3)_{0.05} \cdot 2H_2O$).

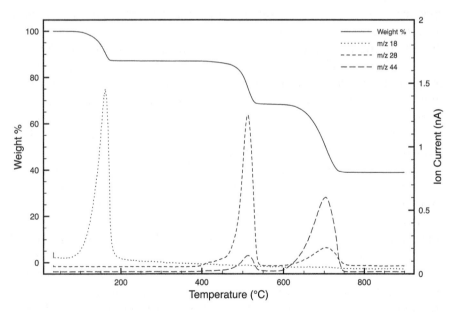

Figure 10.2 TGA plot shown superimposed with MS extracted ions at m/z 18, 28 and 44 representing H_2O, CO and CO_2 respectively (data provided by TA Instruments Waters L.L.C. Used with permission).

The mass spectrometry allowed confirmation of two separate water loss steps at distinct temperatures (350 and 550 K), assigned to loss of lattice water (dehydration) and loss of water from the main brucite type layers during the recombination of hydroxyl groups (dehydroxylation) to form an amorphous product. The MS also confirmed a third high-temperature mass loss to be CO_2 (m/z 44) due to decomposition of a carbonate group. With the increasing interest in gas storage systems such as metal organic frameworks, STA-MS offers a powerful integrated tool for interrogation of these processes. Korablov *et al.* reported an application of the combined approach to examine the gas uptake and release kinetics for putative hydrogen storage systems.[54]

10.3.2 Thermal Analysis-infrared

Adsorption of specific frequencies of infrared radiation induces transitions between vibrational energy levels of molecules.[25] The resulting spectra provide specific information of the identity of the molecular species (and relative amounts from appropriate integration of the spectral envelope) and, when coupled to thermal analytical techniques, provide a powerful hyphenated approach for the chemical speciation and quantification of molecular systems. The absorption process is governed by selection rules that relate to a change in the dipole moment of the molecule[55] and hence non-polar molecules such as homonuclear diatomic gases (*e.g.* molecular oxygen or nitrogen) are not detected and it does not readily distinguish between homologous non-polar series *e.g.* *n*-alkanes. IR spectroscopy can be deployed for solid, liquid and gas samples, and its deployment for EGA from thermogravimetric experiments is perhaps its most widely known application in thermal analysis.

Interfacing a gas sampling cell for infrared spectroscopy is somewhat more straightforward than MS hyphenation (Section 10.3.1), as there is little or no pressure change required between the thermal instrument and the IR gas analyser. The gas cells used generally have relatively long measuring path lengths to enhance the sensitivity of the IR absorption experiment[56] and are attached to a heated transfer line from the thermal analyser. Again, short, inert, heated transfer lines are preferred to minimise condensation, reaction and bulk transfer effects in the eluent. The gas is continuously scanned as it passes through the cell and, in the case of routinely deployed Fourier transform infrared (FTIR) spectrometers, this occurs rapidly and comfortably within the timescales of thermal events. It is worth

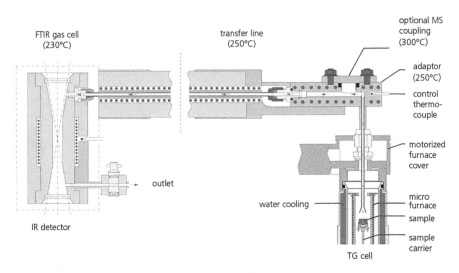

FTIR gas cell (230°C)

transfer line (250°C)

optional MS coupling (300°C)

adaptor (250°C)

control thermo-couple

motorized furnace cover

outlet

water cooling

micro furnace

sample

sample carrier

IR detector

TG cell

Figure 10.3 Schematic figure of TG-FTIR coupling showing integration of heated adaptor, transfer line and FT-IR gas cell (NETZSCH Gerätebau, used with permission).

noting that dispersive infrared spectrometers are rarely used in modern instrumentation.[56] A schematic representation of a typical arrangement for EGA analysis with hyphenated TG-FTIR is shown in Figure 10.3.

TG-FTIR instrumentation has been extended with further hyphenation to MS.[18,45,56,57] As the interface to MS requires minimal sampling of the evolved gas, the remaining eluate can be passed to the FTIR gas cell through a heated transfer line. Such systems offer comprehensive speciation of the evolved gas, *i.e.* FTIR silent species are detected in the MS system. A recent application of this extended hyphenation has been demonstrated for the thermal degradation of a poly(ε-caprolactam)-based polymer.[44] Similarly, STA systems have been adapted for simultaneous FTIR and MS hyphenation.[3,58]

The hyphenation of FTIR with TA for the monitoring of evolved gases was first accomplished in the late 1960s by Kiss,[59] confirming evolution of water and ammonia from their characteristic FTIR spectra. The growth of TG-FTIR has been steady and the technique remains less widely applied than TG-MS. Nevertheless, its utility is considerable and this can be illustrated with the thermal decomposition of calcium oxalate dihydrate (Figure 10.4). The superposition of mass loss data from the thermobalance can be compared with the overall integrated intensity of the IR absorption—available as a Gram–Schmidt plot derived from vector analysis of the acquired

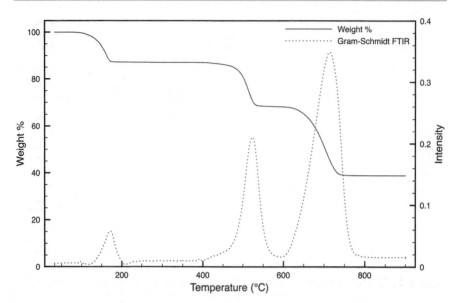

Figure 10.4   TGA-FTIR of calcium oxalate dihydrate heated from ambient to 900 °C at 10 °C min⁻¹ showing mass loss data superimposed with a Gram–Schmidt FTIR intensity plot (data provided by TA Instruments Waters L.L.C. Used with permission).

FITR interferograms.[60] In essence, this approach allows correlation of mass changes with events in the FTIR detector and, in this example, the successive losses of water, CO and CO_2 are apparent. Selecting the interferograms at the appropriate temperatures allows, after Fourier transformation, molecular identification from the FTIR spectrum (Figure 10.5): at 150 °C, the spectrum is consistent with water vapour; at 500 °C, the spectrum shows both CO (apparent as the 'doublet' centred *ca.* 2100 cm⁻¹) and CO_2; at 700 °C, the spectrum is consistent with CO_2.

Simultaneous FTIR and MS hyphenated to simultaneous thermal analysis techniques (STA) is also available. This instrument suite becomes very powerful through providing simultaneous mass loss and heat flow data, in combination with mass spectral and spectroscopic data obtained on the evolved gases. The power of this approach is demonstrated by Jiao *et al.*[22] in deducing the decomposition mechanism of the cyclotrimethylene trinitramine and ammonium perchlorate reaction. In this study, reaction kinetics were derived from combined analysis of TGA/DSC data and the associated mechanistic pathway was validated by molecular speciation from MS and FTIR of the evolved gases.

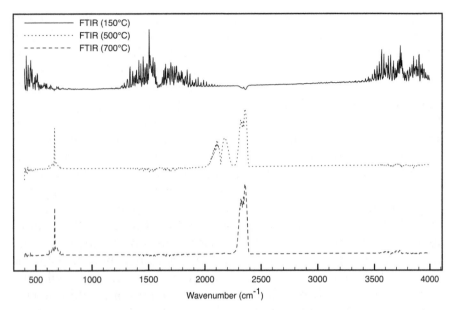

Figure 10.5 FTIR spectra at 150, 500 and 700 °C from the TG-FTIR analysis of the thermal decomposition of calcium oxalate dihydrate. FTIR spectra correspond to water, CO and CO_2 loss (data provided by TA Instruments Waters L.L.C. Used with permission).

10.3.3 Quantitative Analysis using EGA

Both MS and FTIR provide inherently quantitative data on the samples being analysed but calibration of total ion currents for MS and Gram–Schmidt plots for FTIR as the responses in the respective detectors are molecular specific. Systems can be calibrated using a series of quantitatively-prepared mixed gases of varying concentrations, though this can be time-consuming and expensive. More recently, approaches termed 'pulsed techniques' have emerged where injections of specific amounts of inert gases are injected into the purge gas stream in order to calibrate the spectroscopic signals. Such approaches have been used to good effect by Maciejewski *et al.*,[61,62] and have recently been critically compared with traditional calibration approaches in well established reference systems.[63]

10.4 *In Situ* Spectroscopic Analysis

Many physicochemical changes to the nature of a material can complicate interpretation of the thermal trace. These include phase changes, fusion, decomposition reactions and in-cell events such as sintering, separation, foaming and bubbling, and creep of melts that cause inhomogeneity in the sample under study. Such phenomena have provided the impetus for incorporating optical and, more recently, spectroscopic probes into the thermal analyser to provide additional corroboration of events in the thermoanalytical cell. Thermomicroscopy is a well established and described field and, by microscopic interrogation of sample sites, confirmation of melt transitions and/or decompositions, and physical state changes accompanied by volume or colour changes can be readily observed.[64,65] Therefore, this account will concentrate on deployment of spectroscopic methods to in-cell analysis, and their application to the condensed phase materials contained within the cell. It is most appropriate to describe these approaches as simultaneous as they provide information from the sample *in situ* rather than through the sequential analytical train of hyphenated techniques. Mirabella presented what is widely regarded as the first account of combining DSC with FTIR in 1986[66] and an increasing number of accounts have detailed *in situ* analysis of DSC with FTIR,[67–69] Raman[70–72] and near infrared spectroscopy (NIR)[73] as well as simple optical observations *via* white light microscopy.[74] While DSC is the most common thermal approach for in-cell spectroscopic interrogation, Vora *et al.*[75] and Columbano *et al.*[76] describe integration of NIR probes to dynamic vapour sorption instrumentation (DVS-NIR) to study pharmaceutical solid form. DSC-FTIR is most prevalent here but the DSC-Raman method is emerging as an alternative approach. It is these two techniques that are reviewed in the following sections.

10.4.1 Vibrational Spectroscopy

Vibrational state changes are molecular-specific and can be induced through two principal mechanisms: (i) absorption, which forms the basis of FTIR spectroscopy (see Section 10.3.2) and (ii) inelastic scattering of incident radiation that causes a shift in the frequency of the radiation that corresponds with the energy difference of the vibrational states of the molecular species—Raman scattering—which forms the basis of Raman spectroscopy. Combining DSC systems with these spectroscopic methods essentially relies on two approaches: (i) placing micro-DSC instruments (such as the Mettler FP-84 and Linkam DSC600) within the spectrometer or (ii) deploying fibre optic probes into DSC laboratory instrumentation through appropriate windowing of the thermal cell.

In the FTIR experiment, attenuation of the IR radiation by the sample is measured with (i) transmission or (ii) reflectance of the incident beam. For FTIR to be operated in transmission mode, it is necessary to mount the sample in the micro-DSC in a spectroscopically transparent substrate to allow the passage of the energy from source to detector.[77,78] In this configuration, samples are presented to the DSC in the form of either: (i) pressed potassium bromide/chloride (KBr, KCl) discs incorporating the analyte(s); (ii) as a thin film or solid pressed between two pre-formed KBr discs or (iii) in pre-formed KBr sample pans. However, the mismatch of sensitivity of the two techniques presents a significant problem. To avoid saturation of the FTIR detector, relatively small amounts of analyte are used that can lead to poor sensitivity for the DSC experiment. Furthermore, for accurate temperature measurement over the time-period of a spectral acquisition, slow heating rates (typically 2–5 °C) have to be used, again affecting DSC sensitivity. Mirabella *et al.*[66] put forward an ingenious approach to alleviate this problem, whereby a larger sample is presented as the reference in the DSC. This reverses the conventional heat flow signals and increases the sensitivity of the DSC experiment (with a concurrent reduction in resolution) whilst allowing a sample of appropriate size to be incorporated into the sample pan to yield good quality FTIR data without detector saturation. The alternative to transmission is to acquire spectral data in reflectance mode. The DSC setup is modified to allow infrared radiation to impinge on the sample, either with an open pan or with an optically transparent pan lid. The inherently weak reflectance spectrum in the mid-infrared and the concomitant challenges that long FTIR spectral acquisition might bring, has limited such applications to NIR spectroscopy.[67-69]

In contrast, Raman spectroscopy shows considerable promise with this approach. Early studies utilised FT-Raman instrumentation to evaluate polymorphic transitions in organic and inorganic materials.[71,72,78] More recently, dispersive portable Raman instrumentation has been deployed to study organic species.[79-81] With the relatively long acquisition time required for Raman spectra, especially for FT-Raman, the combination appears to be most commonly deployed with isothermal-step DSC experiments. Gaisford *et al.* make use of this technique to hyphenate a dispersive Raman spectrometer (Renishaw RX210 RIAS) to a conventional DSC (TA Instruments Q2000) in their studies on multi-component pharmaceutical product analogues—the co-crystals of benzoic acid and isonicotinamide.[70] This system provides a good illustration of the potential of

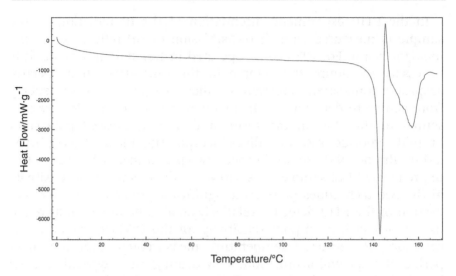

Figure 10.6 DSC trace showing the transition between the 2 : 1 to 1 : 1 co-crystal of benzoic acid–isonicotinamide.

DSC-Raman for the study of complex solid state systems such as this. The instruments were unmodified with the exception of the lidding arrangement to allow the fibre-optic probe access to the material in the pan through an optically transparent accessory and the system was studied between 0 and 170 °C at a heating rate of 10 °C min^{-1}. The polymorphic transition between the 2 : 1 and 1 : 1 co-crystal forms (benzoic acid–isonicotinamide) is clearly evident in the thermogram (Figure 10.6) and spectra (Figure 10.7) are consistent with fully characterised reference samples.

DSC-Raman clearly has potential for wider application and commercial instrumentation has recently become available (Perkin-Elmer's RamanStation™ and RamanFlex™ products).

10.5 Summary

As the capabilities of thermal instrumentation progress and the appetite for innovation of TA platforms continues to develop as the research communities move to interrogate ever more complex materials, the future for hyphenated and combined instrumentation appears to be strong. The convenience of being able to obtain multiple measurements on the same materials conducted with the same subsamples and thermal history is now established, and the increasing array of analytical and structural information that can be

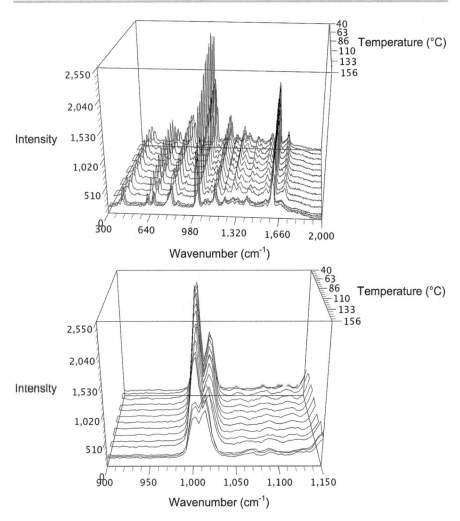

Figure 10.7 Raman spectra acquired over the temperature range 40 to 156 °C showing the transition between the 2:1 to 1:1 co-crystal of benzoic acid–isonicotinamide (top) and spectral expansion between 900 and 1150 cm^{-1} (bottom).

accessed offers a tremendous flexible resource for the materials science community. Hardware developments are likely to benefit from increased miniaturisation of components and increasing automation of operation. More effective integration of data analysis and interpretation, making full use of the parallel information and integrating energetic and structural information, remains a major opportunity in the future development of these combined and hyphenated thermal techniques.

References

1. S. Gaisford and A. B. M. Buanz, *J. Therm. Anal. Calorim.*, 2011, **106**, 221.
2. F. L. Lopez, G. C. Shearman, S. Gaisford and G. R. Williams, *Mol. Pharm.*, 2014, **11**, 4327.
3. P. Rajeshwari and T. K. Dey, *J. Therm. Anal. Calorim.*, 2014, **118**(3), 1531.
4. M. A. Zaman, G. P. Martin, G. D. Rees and P. G. Royall, *Thermochim. Acta*, 2004, **417**, 251.
5. L. P. Esteves, I. Lukosiute and J. Cesniene, *J. Therm. Anal. Calorim.*, 2014, **118**, 1385.
6. H. Liu, T. Chen, X. Zou, C. Qing and R. L. Frost, *Thermochim. Acta*, 2013, **568**, 115.
7. M. Arsenovic, L. Pezo, L. Mancic and Z. Radojevic, *Thermochim. Acta*, 2014, **580**, 38.
8. P. Cortes, I. Fraga, Y. Calventus, F. Roman, J. M. Hutchinson and F. Ferrando, *Materials*, 2014, **7**, 1830.
9. E. E. Sigstad, F. I. Schabes and F. Tejerina, *Thermochim. Acta*, 2013, **569**, 139.
10. R. L. Frost and S. J. Palmer, *Thermochim. Acta*, 2011, **521**, 121.
11. I. Wilinska and B. Pacewska, *J. Therm. Anal. Calorim.*, 2014, **116**, 689.
12. I. O. Ali, *J. Therm. Anal. Calorim.*, 2014, **116**, 805.
13. J. Lopez-Beceiro, S. A. Fontenot, C. Gracia-Fernandez, R. Artiaga and R. Chartoff, *J. Appl. Polym. Sci.*, 2014, **131**, 40670/1.
14. K. Katoh, N. Saeki, E. Higashi, Y. Hirose, M. Sugimoto and N. Katsuyuki, *J. Therm. Anal. Calorim.*, 2013, **113**, 1275.
15. J. L. Heidbrink, J. Li, W. Pan, J. L. Gooding, S. Aubuchon, J. Foreman and C. J. Lundgren, *Thermochim. Acta.*, 1996, **284**, 241.
16. P. A. Barnes, *Anal. Chim. Acta*, 1996, **323**, 323.
17. S. Vecchio, L. Cerretani, A. Bendini and E. Chiavaro, *J. Agric. Food Chem.*, 2009, **57**, 4793.
18. R. L. Frost, V. Vagvoelgyi, S. J. Palmer, J. Kristof and E. Horvath, *J. Colloid Interface Sci.*, 2008, **318**, 302.
19. R. Artiaga, S. Naya, A. García, F. Barbadillo and L. García, *Thermochim. Acta*, 2005, **428**, 137.
20. F. M. Etzler and J. J. Conners, *Thermochim. Acta*, 1991, **189**, 185.
21. A. C. Draye, O. Persenaire, J. Brožek, J. Roda, T. Košek and P. Dubois, *Polymer*, 2001, **42**, 8325.

22. Q. J. Jiao, Y. L. Zhu, J. C. Xing, H. Ren and H. Huang, *J. Therm. Anal. Calorim.*, 2014, **116**, 1125.
23. S. Biamino and C. Badini, *J. Eur. Ceram. Soc.*, 2003, **24**, 3021.
24. R. O. Grishchenko, A. L. Emelina and P. Y. Makarov, *Thermochim. Acta*, 2013, **570**, 74.
25. IUPAC Analytical Chemistry Division, in *Compendium of Analytical Nomenclature (The IUPAC 'Orange Book')*, ed. J. Inczédy, T. Lengyel and A. M. Ure, Blackwell Science, Ltd., Oxford, UK, 3rd edn, 1998.
26. P. Tsytsik, J. Czech, R. Carleer, G. Reggers and A. Buekens, *Chemosphere*, 2008, **73**, 113.
27. P. A. Barnes and G. Stephenson, *Anal. Proc.*, 1981, **18**, 538.
28. R. T. Medeiros, M. I. G. Leles, F. Costa Dias, R. Kunert and N. R. Antoniosi Filho, *J. Therm. Anal. Calorim.*, 2008, **91**, 225.
29. W. W. Wendlandt and T. M. Southern, *Anal. Chim. Acta*, 1965, **32**, 405.
30. F. Zitomer, *Anal. Chem.*, 1968, **40**, 1091.
31. R. Giovanoli and H. G. Weidemann, *Helv. Chim. Acta.*, 1968, **51**, 1134.
32. E. Baumgartner and E. Nachbaur, *Thermochim. Acta*, 1977, **19**, 3.
33. W. Scwaneback and H. W. Wenz, *Anal. Chem.*, 1988, **331**, 61.
34. K. W. Samlldon, R. E. Ardrey and L. R. Mullings, *Anal. Chim. Acta*, 1979, **107**, 327.
35. E. Kaisersberger, *Int. J. Mass Specrom. Ion Phys.*, 1983, **46**, 155.
36. K. H. Ohrbach, G. Radhoff and A. Kettrup, *Int. J. Mass Specrom. Ion Phys.*, 1983, **47**, 59.
37. E. Clarke, *Thermochim. Acta*, 1981, **51**, 7.
38. E. L. Charsley, N. J. Manning and S. B. Warrington, *J. Calorim., Anal. Therm. Thermodyn. Chim.*, 1986, **17**, 429.
39. E. L. Charsley, N. J. Manning and S. B. Warrington, *Thermochim. Acta*, 1987, **114**, 47.
40. K. G. H. Raemaekers and J. C. J. Bart, *Thermochim. Acta*, 1997, **295**, 1.
41. V. N. Emel'yanenko, G. Boeck, S. P. Verevkin and R. Ludwig, *Chem. – Eur. J.*, 2014, **20**, 11640.
42. T. Tsuneto, I. Murasawa, M. Nagata and Y. Kubota, *J. Anal. Appl. Pyrolysis*, 1995, **33**, 139.
43. D. Dollimore, G. A. Gamlen and T. J. Taylor, *Thermochim. Acta*, 1984, **75**, 59.
44. A. C. Draye, O. Persenaire, J. Brožek, J. Roda, T. Košek and P. Dubois, *Polymer*, 2001, **42**, 8325.
45. O. Persenaire, M. Alexandre, P. Degée and P. Dubois, *Biomacromolecules*, 2001, **2**, 288.

46. R. M. M. DosSantoz, R. G. L. Gonçalves, V. R. L. Constantino, L. M. daCosta, L. H. M. daSilva, J. Tronto and F. G. Pinto, *Appl. Clay Sci.*, 2013, **80–81**, 189.

47. B. C. Miranda, R. J. Chimentao, J. B. O. Santos, F. Gispert-Guirado, J. Llorca, F. Medina, F. L. Bonillo and J. E. Sueiras, *Appl. Catal., B*, 2014, **147**, 464.

48. K. Yoo and P. G. Smirniotis, *Appl. Catal., A*, 2003, **246**, 243.

49. P. Tsytsik, J. Czech, R. Carleer, G. Reggers and A. Buekens, *Chemosphere*, 2008, **73**, 113.

50. R. Mellaerts, K. Houthoofd, K. Elen, H. Chen, M. vanSpeybroeck, J. vanHumbeeck, P. Augustijns, J. Mullens, G. vandenMooter and J. A. Martens, *Microporous Mesoporous Mater.*, 2010, **130**, 154.

51. M. G. Mothé, C. G. Mothé, C. H. M. Carvalho and M. C. Khalil de Oliveira, *J. Therm. Anal. Calorim.*, 2014, **117**, 1357.

52. S. Lüftl, V. M. Archodoulaki and S. Seidler, *Polym. Degrad. Stab.*, 2006, **94**, 464.

53. R. O. Grishchenko, A. L. Emelina and P. Y. Makarov, *Thermochim. Acta*, 2013, **570**, 74.

54. D. Korablov, F. Besenbacher and T. R. Jensen, *Int. J. Hydrogen Energy*, 2014, **39**, 9700.

55. B. H. Stuart, *Infrared Sprectroscopy: Fundamentals and Applications*, 2004, John Wiley & Sons Ltd.

56. S. Galvagno, S. Casu, M. Martino, E. DiPalma and S. Portofino, *J. Therm. Anal. Calorim.*, 2007, **88**, 507.

57. W. Pawelec, M. Aubert, R. Pfaendner, H. Hoppe and C. E. Wilén, *Polym. Degrad. Stab.*, 2012, **97**, 948.

58. L. Jiao, G. Xu, Q. Wang, Q. Xu and J. Sun, *Thermochim. Acta*, 2012, **547**, 120.

59. A. B. Kiss, *Acta. Chim. Acad. Sci. Hung.*, 1969, **61**, 207.

60. D. M. Price and S. P. Church, *Thermochim. Acta*, 1997, **294**, 107.

61. M. Maciejewski and A. Baiker, *Thermochim. Acta*, 1997, **295**, 95.

62. M. Maciejewski, C. A. Muller, R. Tschan, W. D. Emmerich and A. Baiker, *Thermochim. Acta.*, 1997, **295**, 167.

63. F. Eigenmann, M. Maciejewski and A. Baiker, *J. Therm. Anal. Calorim.*, 2006, **83**, 321.

64. Y. Leng, 2008, Thermal Analysis, in *Materials Characterisation: Introduction to Microscopic & Spectroscopic Methods*, John Wiley & Sons (Asia), Pte Ltd, Singapore.

65. M. E. Brown, 2001, Thermoptometry, in *Introduction to Thermal Analysis; Techniques and Applications*, Springer, Netherlands.

66. F. M. Mirabella, *Appl. Spectrosc.*, 1986, **40**, 417.

67. H. L. Lin, T. K. Wu and S. Y. Lin, *Thermochim. Acta*, 2014, **575**, 313.

68. S. D. Pandita, L. Wang, R. S. Mahendran, V. R. Machavaram, M. S. Irfan, D. Harris and G. F. Fernando, *Thermochim. Acta*, 2012, **543**, 9.
69. S. L. Wang, Y. C. Wong, W. T. Cheng and S. Y. Lin, *J. Therm. Anal. Calorim.*, 2011, **104**, 261.
70. A. B. M. Buanz, R. Telford, I. J. Scowen and S. Gaisford, *CrystEngComm*, 2013, **15**, 1031.
71. J. C. Sprunt, U. A. Jayasooriya and R. H. Wilson, *Phys. Chem. Chem. Phys.*, 2000, **2**, 4299.
72. M. E. E. Harju and J. Valkonen, *Spectrochim. Acta*, 1991, **47A**, 1395.
73. C. J. deBakker, N. A. St. John and G. A. George, *Polym. Prepr. (Am. Chem. Soc., Div. Polym. Chem.)*, 1992, **33**, 374.
74. P. Kuo, C. Lo and C. Chen, *Polymer*, 2013, **54**, 6654.
75. K. L. Vora, G. Buckton and D. Clapham, *Eur. J. Pharm. Sci.*, 2004, **22**, 97.
76. A. Columbano, G. Buckton and P. Wilkeley, *Int. J. Pharm.*, 2002, **237**, 171.
77. H. L. Lin, G. C. Zhang, Y. Huang and S. Lin, *J. Pharm. Sci.*, 2014, **103**, 2386.
78. H. L. Lin, T. K. Wu and S. Y. Lin, *Thermochim. Acta*, 2014, **575**, 313.
79. J. C. Sprunt and U. A. Jayasooriya, *Appl. Spectrosc.*, 1997, **51**, 1410.
80. H. R. H. Ali, H. G. M. Edwards, M. D. Hargreaves, T. Munshi, I. J. Scowen and R. J. Telford, *Anal. Chim. Acta*, 2008, **620**, 103.
81. H. R. H. Ali, H. G. M. Edwards and I. J. Scowen, *J. Raman Spectrosc.*, 2009, **40**, 887.

11 Sample Controlled Thermal Analysis

G. M. B. Parkes[*a] and E. L. Charsley[b]

[a] Department of Chemical Sciences, University of Huddersfield, Queensgate, Huddersfield HD1 3DH, UK; [b] IPOS, School of Applied Sciences, University of Huddersfield, Queensgate, Huddersfield HD1 3DH, UK
*Email: g.m.b.parkes@hud.ac.uk

11.1 Introduction

In conventional thermal analysis, a sample is subjected to a pre-determined heating programme, typically a linear rising ramp. Although simple in concept, the use of linear heating causes temperature gradients across the sample, arising from the conduction of heat from the external furnace, and concentration gradients, arising from the diffusion of reaction or product gases into or out of the sample. These gradients result in non-uniform reaction conditions throughout the sample leading to loss of resolution of thermally adjacent processes.

To address these problems, a number of alternative approaches to conventional linear heating have been developed where the heating rate is governed by a property of the sample itself.[1] The generic name for this family of techniques is 'sample controlled thermal analysis' (SCTA).[2] The technique was pioneered independently in the 1960s by Rouquerol in France who developed a method for use at reduced pressures called constant rate thermal analysis[3] and by the Paulik brothers in Hungary who patented an approach called

Principles of Thermal Analysis and Calorimetry: 2nd Edition
Edited by Simon Gaisford, Vicky Kett and Peter Haines
© The Royal Society of Chemistry 2016
Published by the Royal Society of Chemistry, www.rsc.org

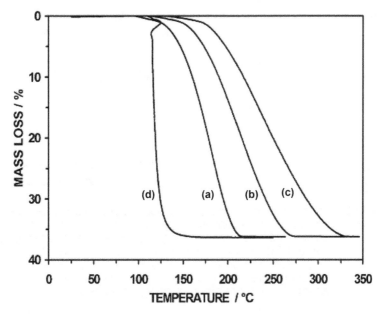

Figure 11.1 Comparison of the mass loss traces as a function of temperature for the decomposition of sodium hydrogen carbonate
using (a) 2 °C min^{-1}; (b) 5 °C min^{-1}; (c) 10 °C min^{-1}; (d) SCTA.[6]

quasi-isothermal quasi-isobaric thermal analysis using their Derivatograph simultaneous TG-DTA apparatus.[4] In a later development,
Sørensen introduced the stepwise isothermal analysis technique.[5]

The advantage of using SCTA is illustrated by comparing a SC-TG
experiment on a 500 mg sample of sodium hydrogen carbonate with
TG experiments carried out at three different heating rates
(Figure 11.1).[6] It is immediately apparent that in the SC-TG experiment, which was carried out under constant rate conditions (see
Section 11.2.1), the majority of the decomposition occurred at a
significantly lower temperature and over a smaller temperature range
than in the linear heating experiments. In addition, because cooling
as well as heating is allowed, the fall in temperature (shown as the
mass loss profile curving back on itself) needed to keep the rate of
decomposition constant gives mechanistic information and is
indicative of a nucleation and growth mechanism (see Section 11.3).

11.2   SCTA Control Techniques

In a conventional thermal analysis system (Figure 11.2a), the heating
rate is pre-determined, while in an SCTA system (Figure 11.2b), the

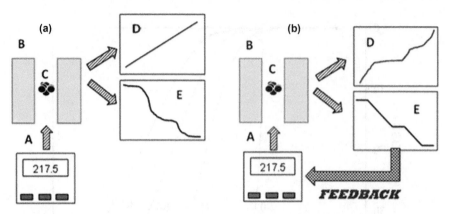

Figure 11.2 Schematic of a general thermal analysis system. Temperature controller (A) supplies power to a furnace (B) which heats sample (C). In a linear heating experiment, (a), the temperature curve (D) is pre-determined while the response of the sample (E) is recorded. In a SCTA experiment, (b), there is feedback between the process and the temperature controller—in effect, the sample dictates how it is heated.

heating rate is governed by a feedback loop dependent on the rate of the reaction or transformation undergone by the sample itself. It is the nature of this feedback that distinguishes between the different types of SCTA.[7] The majority of SCTA techniques reduce the heating rate (to low, zero or even cooling) *during* a process and increase the heating rate *between* processes although other approaches are possible.

In contrast to thermoanalytical profiles obtained using linear heating, those produced under SCTA conditions are often very different in appearance when plotted as a function of either time or temperature.[8] In some cases, SCTA data is easiest to interpret if the process is presented in the form of the extent of reaction, α (Figure 11.3).

11.2.1 Constant/Controlled Rate Thermal Analysis

Following the development of constant rate thermal analysis, Rouquerol *et al.* proposed the term 'controlled' instead of 'constant' to cover cases where the composition of evolved gases changed during a decomposition reaction.[9] Both names are used in the literature and have the same acronym 'CRTA'. In this approach, the sample temperature is altered so as to keep the rate of reaction, or transformation, constant at a pre-set level. This has the effect of reducing both the temperature and pressure gradients in the sample thus eliminating the uncertainties present in linear heating experiments.

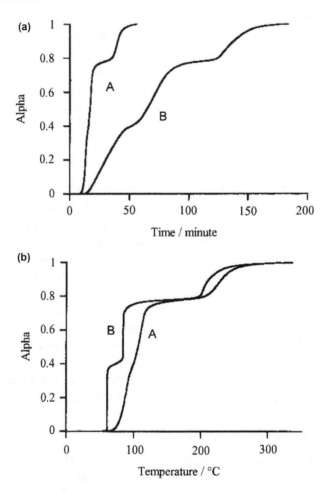

Figure 11.3 Comparison of α traces for the dehydration of copper sulfate pentahydrate using linear heating (curve A) and proportional heating SCTA (curve B). (a) α *vs.* time; (b) α *vs.* temperature. Reprinted with permission from G. M. B. Parkes, P. A. Barnes and E. L. Charsley, *Anal. Chem.*, 1999, **71**, 2482. Copyright 1999 American Chemical Society.[8]

Since CRTA is able to control closely the conditions in the sample, it is the SCTA technique most widely used for kinetic studies (Section 11.3) and as a preparative method (Section 11.4.8).

11.2.2 Stepwise Isothermal Analysis

In stepwise isothermal analysis (SIA), the sample is subjected to a linear temperature programme until the reaction rate exceeds a pre-determined level at which point the heating is stopped. The re-action thereafter proceeds isothermally until the rate falls below the set level whereupon heating is resumed. By using this approach, the reaction is made to take place in several isothermal steps. Although SIA has been successfully applied to both thermogravimetry and the sintering of ceramics,[10] care needs to be taken as it is possible to introduce artefacts in the reaction profiles.[11]

11.2.3 Constrained Rate Techniques

Constrained rate is a term, first used by Reading,[11] that describes a range of techniques, where the temperature is decreased as a function of the reaction rate but no attempt is made to maintain the latter at a constant value. These include the proportional heating technique of Parkes *et al.*,[8] the HiRes™ dynamic rate technique of TA Instruments,[12] the MaxRes technique of Mettler-Toledo[13] and the Super-Res® of Netzsch.[14] There can be considerable variation in the operation of these techniques and whether cooling, rather than just slow heating, is supported. However, the aim of all the constrained rate techniques is to increase resolution in the minimum time. So, while they do not provide the quality of kinetic data that can be obtained using CRTA, the experiments are significantly shorter in duration and hence may be more applicable to industrial studies where time may be at a premium.

11.2.4 Peak Shape Techniques

These techniques use the shape of a derivative TG (DTG) or evolved gas analysis (EGA) peaks in multi-stage reactions to control the heating rate.[8,15] They have the aim of overcoming the limitation of the SCTA techniques described in the preceding sections, which require a pre-selected reaction rate, by allowing the reaction itself to determine the heating programme. In the simplest form, the heating rate is switched to zero when the first reaction peak rate reaches a maximum. The temperature is then held until a peak minimum is achieved when the heating rate is increased again until the next peak maximum. In this way, irrespective of the absolute magnitude of the peaks, the changes in heating rate will occur at the same relative position on each peak.[8]

11.2.5 Comparison of Different SCTA Techniques

A comparison of results obtained using the different SCTA techniques, compared with linear heating, is shown in Figure 11.4 for the

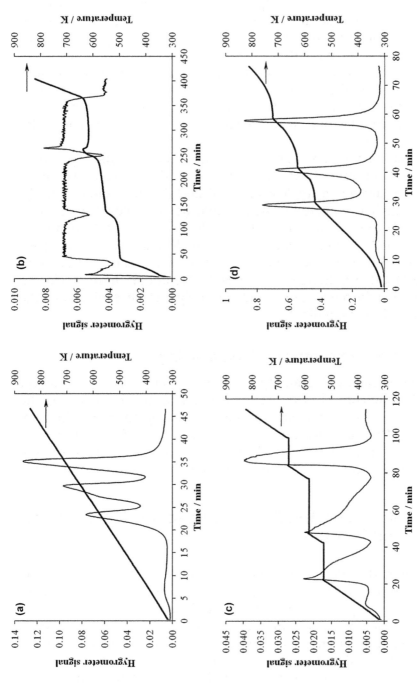

Figure 11.4 Comparison of the water loss traces for the decomposition of a mixture of nickel, magnesium and calcium hydroxides monitored using EGA (hygrometer) under conditions of (a) linear heating; (b) CRTA; (c) SIA; (d) peak shape analysis.

decomposition of a physical mixture of nickel, magnesium and calcium hydroxides. The decomposition was followed by EGA using a hygrometer and under linear heating showed a small initial evolution of water followed by three overlapping dehydroxylation peaks (Figure 11.4a). In the CRTA experiment, each decomposition stage was made to proceed at a low rate and took approximately ten times longer than in the linear heating experiment (Figure 11.4b). The temperature profile shows that each reaction occurred over a narrower range and with a greater separation between stages than under linear heating.

For clarity, the experimental conditions for the SIA experiment (Figure 11.4c) were chosen so that each of the dehydroxylation reactions occurred in a single isothermal step although, more typically, several smaller temperature steps would be observed. Using a 'peak shape' SCTA technique (Figure 11.4d), the three EGA peaks retain the approximately Gaussian shape observed during linear heating experiments but are now more clearly resolved. It can be seen from Figure 11.4 that although all three SCTG techniques have provided a reduction in the temperature gradients, only CRTA has minimised the concentration gradients.[16]

11.2.6   Gas Concentration Programming

Gas concentration programming is an extension of the SCTA concept in which the control of the reaction rate is achieved by varying the concentration of a reactant gas rather than by changing the temperature. This approach can be useful in the study of exothermic gas–solid reactions, such as the oxidation of carbons,[17] where it may not be possible to control the reaction by programming the temperature.

11.3   SCTA and the Study of Reaction Mechanisms and Kinetics

Most thermal analysis kinetic determinations use data from dynamic heating experiments. However, the benefits of CRTA, where non-uniformity of reaction conditions produced by temperature and pressure gradients are minimised, means that the reliability of kinetic data achievable from even a single experiment is greatly increased compared with linear heating methods.[18]

To aid identification of the correct kinetic model, theoretical CRTA curves of α against temperature have been derived.[19–21] Not only can

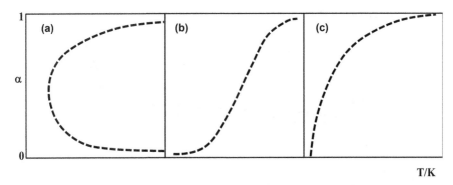

Figure 11.5 Theoretical α *vs.* temperature traces for CRTA experiments. (a) Nucleation and growth, (b) diffusion and (c) phase boundary.

the main kinetic model types, nucleation and growth/autocatalysis, and diffusion and phase boundary be identified (Figure 11.5) but it is also possible to discern between different models of a given type.

F. and J. Rouquerol proposed a modification to CRTA called the 'rate-jump' technique, which allows the determination of the activation energy of a process irrespective of any knowledge of the kinetic model.[22] In this technique, the reaction rate is periodically switched between two values, C_1 and C_2, and the corresponding temperatures, T_1 and T_2, just before and after each 'jump' is recorded. Assuming that the two rates, C_1 and C_2, follow the same mechanism and that α changes negligibly during the course of a rate jump, then the apparent activation energy (E_a) can be given by:

$$E_a = \left(\frac{RT_1T_2}{T_2 - T_1} \right) \times \ln\left(\frac{C_2}{C_1} \right) \tag{11.1}$$

If the overall rates are low, then a number of rate jumps can be performed in a single experiment thus allowing the activation energy to be determined throughout the process.

Rouquerol *et al.* used the rate-jump technique in a study of the effect of water vapour pressure on the dehydroxylation of kaolinite clay.[23] Figure 11.6 shows the temperature profile for a rate-jump experiment performed under a water vapour pressure of 500 Pa. Over the 140 h experiment, 23 separate rate jumps were obtained. From calculations from each rate jump, the authors showed that E_a was constant to $\alpha = 0.84$, with an average of 190 kJ mol^{-1}.

CRTA has been applied to study the kinetics of a wide range of solid-state reactions. The benefits of using this approach have been discussed in a number of useful reviews.[7,24,25]

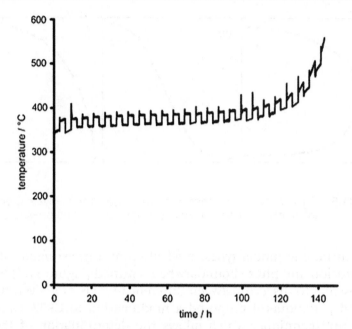

Figure 11.6 Kaolinite dehydroxylation carried out under CRTA conditions using the rate-jump method. The steps in the temperature curve correspond to the switch in decomposition rates. Reprinted from K. Nahdi, P. Llewellyn, F. Rouquerol, J. Rouquerol, N. K. Ariguib and M. T. Ayedi, *Thermochim. Acta*, 2002, **390**, 123. Copyright (2002), with permission from Elsevier.[23]

11.4   SCTA Instrumentation and Applications

This section highlights the different thermal analysis methods to which SCTA techniques have been applied. Examples have been chosen to illustrate some of the benefits that may be obtained using these techniques.

11.4.1 Thermogravimetry

Thermogravimetry is now the most widely used thermal method for SCTA studies. The majority of the early work in this field was carried out by the Paulik brothers and their co-workers and by users of the Derivatograph and the work has been the subject of an extensive review by Paulik.[26] The use of the technique was considerably extended from the 1990s onwards by the introduction of SCTA software by the major thermal analysis equipment manufacturers. The range of commercial SCTA techniques currently available is summarised in Table 11.1.

Table 11.1 Commercially available SC-TG programmes.

Company	Technique	Reference
Mettler-Toledo	MaxRes	13
Netzsch[a]	Super-Res®	14
Perkin Elmer	AutoStepwise (SIA)	27
Rigaku	Constant Rate	28
SETARAM[a]	Constant Rate	29
TA Instruments	HiRes™ (Dynamic Rate, Constant Rate)	12

[a]Programmes also available for dilatometry.

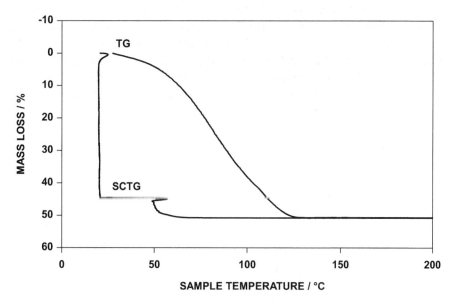

Figure 11.7 TG and SC-TG traces for the dehydration of strontium hydroxide octahydrate.[30]

An illustration of the increased resolution that can be obtained under SC-TG conditions is shown by a study on the dehydration of strontium hydroxide octahydrate.[30] TG experiments showed a single stage mass loss over the region ambient to 130 °C (Figure 11.7). This corresponded to the loss of 8 H_2O after allowing for the presence of 7% of strontium carbonate in the sample. The SC-TG curve showed that the sample was initially cooled slightly to about 20 °C and held at this temperature until 7 H_2O had been removed. The sample was then heated to 50 °C, at which temperature the remaining water was lost. The shape of the mass loss curve for the decomposition of the monohydrate is characteristic of that attributed to a mechanism involving nucleation and growth of nuclei (see Section 11.3).

The higher resolution obtained using constrained rate techniques, compared with linear heating, has resulted in their application to the study of overlapping reactions and to the analysis of multi-component systems, often using the DTG curves. This is illustrated by the work of Fernández-Berridi *et al.*[31] who have applied HiRes™ TGA in the dynamic rate mode to study the natural rubber (NR)–styrene butadiene elastomer (SBR) blends used in tyre rubbers.

The DTG curves obtained for a range of blends are shown in Figure 11.8. Two peaks were observed, the first around 350 °C due to the decomposition of NR and the second in the region of 420 °C to the SBR. A calibration curve was produced, using compositions containing known amounts of SBR and NR, by plotting H_{NR}/H_{SBR} against %NR, where H_{NR} and H_{SBR} are the respective heights of the DTG peaks for NR and SBR. Using this curve, the authors were able to obtain accurate values for the NR and SBR components in a number of tyre rubbers.

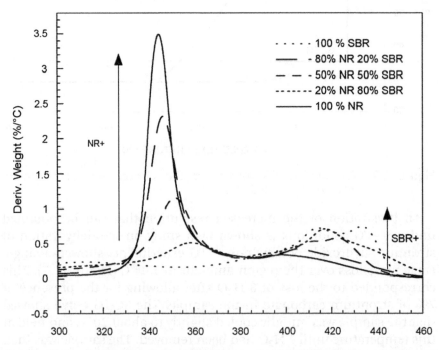

Figure 11.8 Constrained rate DTG traces for NR/SBR blends of different elastomer compositions.
Reprinted from M. J. Fernández-Berridi, N. González, A. Mugica and C. Bernicot, *Thermochim. Acta*, 2006, **444**, 65. Copyright (2006), with permission from Elsevier.[31]

Other constrained rate studies include poly (vinyl chloride) resins,[32] ethylene propylene diene monomer elastomers,[33] thermoplastic wood components,[34] diesel fuel additives,[35] organically modified clays[36] and Portland cement–silica blends.[37]

The introduction of humidity generators to enable measurements to be performed under controlled humidity conditions is of significant benefit when carrying out SCTA experiments. This is illustrated by a study on the thermal decomposition of zinc acetylacetonate as a means of preparing pure zinc oxide.[38] The SC-TG curves obtained in dry and wet helium $(P_{H_2O} = 11.9\,\text{kPa})$ are shown in Figure 11.9.

In dry conditions, a two-stage mass loss was observed corresponding to dehydration followed by complete sublimation of the anhydrous zinc acetylacetonate above 100 °C. In contrast, under conditions of controlled humidity, a single step reaction was observed below 100 °C, leading to the formation of crystalline zinc oxide by the reaction:

$$Zn(CH_3COHCOCH_3)_2 \cdot H_2O \rightarrow ZnO + 2CH_3CH_2COCH_3$$

The characteristic shape of the SC-TG curve indicates that the reaction took place by a nucleation and growth mechanism.

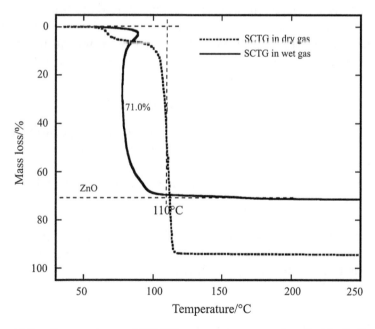

Figure 11.9 Comparison of SC-TG mass loss traces for $C_{10}H_{14}O_4Zn \cdot H_2O$ in a dry helium atmosphere and in a helium atmosphere of controlled humidity.
With kind permission from Springer Science and Business Media.[38]

11.4.2   Evolved Gas Studies

11.4.2.1   *Evolved Gas Detection*

Much of the pioneering work of Rouquerol in the field of SCTA used evolved gas detection (EGD) where a pressure gauge formed a simple but sensitive means of studying dehydration, dehydroxylation and other decomposition reactions.[39] By continuously evacuating the sample cell, using a vacuum pump and a diaphragm constriction, experiments can be carried out at an accurately defined pressure. The system allows relatively large sample masses to be used, enabling preparative-scale experiments to be performed (see Section 11.4.8), and has also been successfully applied to low temperature desorption studies.[40]

This approach is illustrated by a study of the dehydration of gypsum, $CaSO_4 \cdot 2H_2O$,[41] where experiments were carried out under three different residual water vapour pressures, namely 1, 500 and 900 Pa. Since the dehydration was performed at a constant rate, and water was the only product evolved, the length of each dehydration step was directly related to the corresponding mass loss.

Under the two lower pressures, only a single dehydration step was observed and this is illustrated in Figure 11.10a for an experiment at 500 Pa. The overall mass loss, determined by weighing the reaction vessel at the beginning and end of the experiment, was 20.9%, which corresponded to the reaction:

$$CaSO_4 \cdot 2H_2O \rightarrow CaSO_4 + 2H_2O$$

Increasing the pressure to 900 Pa caused the dehydration to take place in two stages (Figure 11.10b) with a similar overall mass loss. The ratio of steps AC and CE was 3 : 1 corresponding to the formation of the calcium sulfate hemihydrate as a reaction intermediate:

$$CaSO_4 \cdot 2H_2O \rightarrow CaSO_4 \cdot 0.5H_2O + 1.5H_2O$$

These experiments demonstrate the advantage of the CR-EGD technique in precisely controlling the pressure of an evolved gas under vacuum conditions, which would not be possible under linear heating conditions.

11.4.2.2   *Specific Gas Detectors*

Use of a specific gas detector allows an evolved gas to be identified giving additional chemical information about the process. For instance, Fesenko *et al.*[16] used a hygrometer to monitor the

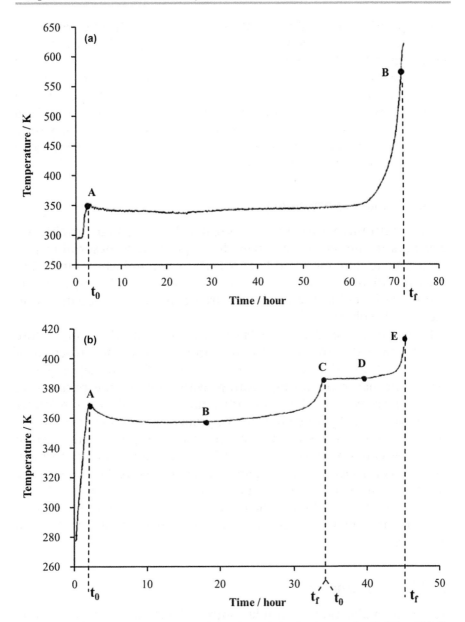

Figure 11.10 CR-EGD traces for the dehydration of $CaSO_4 \cdot 2H_2O$. (a) $P_{H_2O} = 500$ Pa; (b) $P_{H_2O} = 900$ Pa.
Reprinted from E. Badens, P. Llewellyn, J. M. Fulconis, C. Jourdan, S. Veesler, R. Boistelle and F. Rouquerol, *J. Solid State Chem.*, 1998, **139**, 37. Copyright (1998), with permission from Elsevier.[41]

decomposition of metal hydroxides (see Figure 11.4). In cases where more than a single gas was evolved, multiple specific gas detectors have proved beneficial. Reading and Rouquerol used a system with IR gas analysers to monitor the water and carbon dioxide evolution during the decomposition of basic metal carbonates.[42] By performing a CRTA experiment where the rate of CO_2 evolution was maintained constant while the rate of water evolution was monitored, they were able to demonstrate that the two gases were not evolved at the same rate throughout the decomposition.

11.4.2.3 Mass Spectrometry

Mass spectrometry (MS) provides greater flexibility than specific gas detectors in monitoring complex decomposition or desorption processes. Rouquerol modified his CR-EGD system to incorporate MS in the early 1990s producing a system that allowed CRTA experiments controlled by either the total pressure or the partial pressure of an individual evolved gas.[43]

Barnes *et al.* used a SCTA-MS system to study the temperature programmed desorption and temperature programmed reaction of propan-2-ol adsorbed on an acidified clay.[44] Two overlapping processes are observed during heating, with some of the propan-2-ol (major ion at $m/z = 45$) being desorbed unreacted while the remainder is catalytically dehydrated to produce propene (major ion at $m/z = 39$). In an attempt to determine whether the desorption and reaction processes could be resolved, two CRTA experiments were performed, the first controlled by the propan-2-ol signal (Figure 11.11a) and the second by propene (Figure 11.11b). Although it was not possible to obtain complete resolution, the temperature profiles in the two CRTA experiments show the mechanistic differences between the two processes.

11.4.3 Dilatometry

The main application of sample controlled techniques to dilatometry has been carried out by Sørensen who used his SIA technique to study the sintering of ceramics.[45,46] In these measurements, which he called stepwise isothermal dilatometry (SID), the temperature was controlled by the change in the sample length corresponding to shrinkage due to densification during sintering. The technique enabled the E_a values and the controlling mechanisms to be determined for the initial stages of sintering. It should be noted that experiments

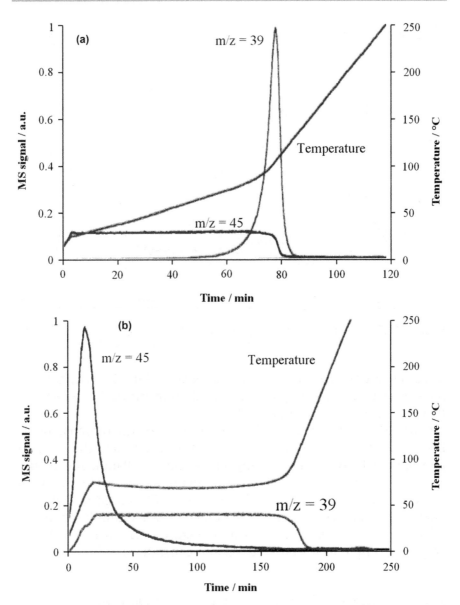

Figure 11.11 CR-EGA traces for the conversion of propan-2-ol ($m/z = 45$)
to propene ($m/z = 39$) on an acid clay as monitored using MS.
(a) Control on propanol; (b) control on propene.
Reprinted from P. A. Barnes, G. M. B. Parkes, D. R. Brown and
E. L. Charsley, *Thermochim. Acta*, 1995, **269–270**, 665.
Copyright (1995), with permission from Elsevier.[44]

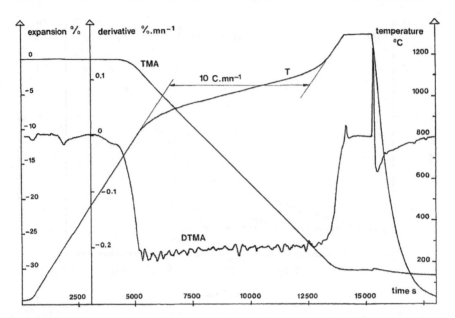

Figure 11.12 Thermodilatometry experiment showing the sintering of a hydroxyapatite sample under constant rate conditions.[50]

in some recent SID papers using Sørensen's kinetic treatment approach have been carried out with fixed temperature increments rather than under sample control.[47,48]

Other SCTA programmes have been used for dilatometry. In the first SC-dilatometry study, Palmour and Johnson used the constant rate approach to study sintering of ceramics.[49] Paulik ingeniously adapted the Derivatograph to enable dilatometric measurements to be performed.[26] As noted in Table 11.1, both SETARAM and Netzsch offer SCTA software for their dilatometers. An example of an SC-dilatometry experiment is shown in Figure 11.12 where a hydroxyapatite sample has been sintered at a constant rate.[50]

11.4.4   Differential Thermal Analysis and Differential Scanning Calorimetry

To date, SCTA techniques have been based largely on the measurement of changes in gas concentration or in mass, although the temperature difference method developed by Smith in 1940[51] could be considered as the first SCTA technique. This approach, generally referred to as 'Smith Thermal Analysis', used a differential thermocouple to establish a constant temperature difference between the sample and furnace wall. It is still being employed in a modernised form to study alloy systems.[52]

Paulik and co-workers carried out the first SC-DTA studies by modifying the Derivatograph simultaneous TG-DTA apparatus to enable the rate of the reaction to be controlled by the derivative of the DTA signal.[53] They describe the application of this technique to the study of a range of inorganic samples with the main emphasis on dehydration and decomposition reactions.[54]

Recently, Charsley *et al.* have developed a SC-DSC system based on a heat flux DSC.[55] This new technique has extended the range of reactions that fall within the scope of SCTA, particularly those taking place without a change in mass. This is illustrated in Figure 11.13a, which shows the curing of an epoxy resin carried out using the proportional heating technique. The experiment is compared with a DSC experiment at 5 °C min^{-1} in Figure 11.13b and shows the reduced temperature range over which a reaction can take place under sample controlled conditions. The use of SC-DSC should allow the development of temperature profiles to enable epoxy resin samples to be cured at a selected rate.

One problem with sample controlled techniques is the difficulty of making baseline corrections, *e.g.* for the 'buoyancy' effect in TG experiments, since the baseline for a given experiment is specific to both the sample and to the experimental conditions. This has been addressed by incorporating a facility to record the temperature programme obtained during an SCTA experiment, which can then be replayed during subsequent experiments. This enables a baseline to be determined under the same conditions as for the sample but with an empty crucible or the reaction products.[55]

11.4.5 Thermomicroscopy

A system for carrying out thermomicroscopy under constant rate or stepwise isothermal analysis conditions has been described by Charsley *et al.*[56] The system used a commercial hot stage operating over the range −180 °C to 600 °C using a silver block heater in conjunction with a liquid nitrogen cooling system. The standard temperature programme was modified so that the sample temperature was controlled by the rate of change of the light intensity signal obtained from a photocell fitted into the microscope eyepiece.

Reflected light intensity measurements were performed under white light and depolarised light intensity (DLI) measurements were carried out using transmitted polarised light. The latter

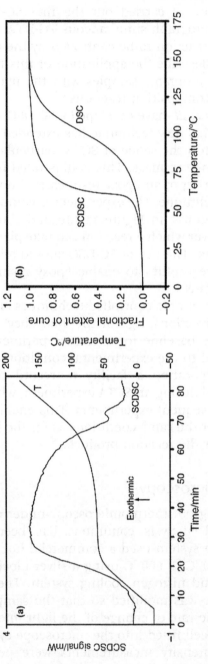

Figure 11.13 (a) Curing of an epoxy resin using SC-DSC; (b) comparison of the fractional extent of cure *versus* temperature traces for DSC and SC-DSC experiments. With kind permission from Springer Science and Business Media.[55]

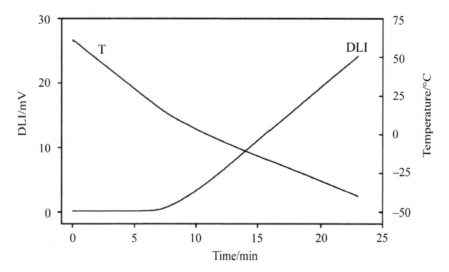

Figure 11.14 Constant rate depolarised light intensity (DLI) trace obtained under cooling conditions for a crude oil.
With kind permission from Springer Science and Business Media.[56]

measurements were made using crossed-polars so that the only light transmitted is due to rotation of the polarised light by the crystalline structure of the sample. Thus any change in the structure on heating will result in a corresponding change in the transmitted light intensity. It is therefore possible to apply the technique to a range of transformations which take place without a change in mass including phase transitions, fusion and recrystallisation reactions.

The technique is illustrated by a study on the behaviour of a crude oil sample on cooling under constant rate conditions (Figure 11.14). As the sample was cooled below room temperature, the DLI signal began to increase due to the formation of crystalline wax particles and the cooling rate was reduced. It can be seen that in order to maintain a constant rate of wax formation, the sample had to be cooled at a linear rate. This suggested that the amount of wax formed was directly related to the temperature and that there were no significant hysteresis effects.

11.4.6 Application of SCTA to Simultaneous Thermal Analysis Methods

In this section, we discuss the application of SCTA techniques to simultaneous thermal methods where more than one thermal measurement is carried out on the same sample at the same time.

It should be noted that the sample control is only applied to one of the thermal methods.

In an early example of this approach, Rouquerol interfaced a vacuum thermobalance to his CR-EGD apparatus.[57] Pérez-Maqueda *et al.* have described a system for studying the decomposition of very stable materials where a quadrupole mass spectrometer was used to control the reaction rate.[58] The system incorporated a vacuum thermobalance thus providing a simultaneous record of the mass changes. For measurements at ambient pressure, Koga and Yamada coupled an EGA system, based on an infrared CO_2 detector and a hygrometer, to a thermobalance. In this instrument, the rate of reaction was controlled by the CO_2 signal.[59]

A powerful option for materials characterisation is the application of SC-TG control to a simultaneous TG-MS or TG-FTIR system. An example of the former approach is that of Arii and Masuda who used a simultaneous SC-TG/MS system to study the thermal decomposition of calcium copper acetate hexahydrate under constant rate conditions.[60] The results of an SC-TG/MS experiment are shown in Figure 11.15a where the SC-TG, temperature and total ion current (TIC) signals are plotted as a function of time. Following the dehydration stage, MS scans during the second decomposition stage in the region of 200 °C revealed that CO, CO_2, CH_3COOH and CH_2 were evolved. The measured mass loss agreed closely with the theoretical value for the decomposition reaction:

$$2CaCu(CH_3CO_2)_4 \rightarrow 2Ca(CH_3CO_2)_2 + Cu_2O + 2CH_2$$
$$+ 2CH_3COOH + CO + CO_2$$

This mechanism was confirmed by X-ray diffraction studies on the reaction intermediate.

The importance of carrying out the decomposition under sample controlled conditions is shown when the SC-TG and TG curves are compared in Figure 11.15(b). It can be seen that the second mass loss stage was markedly larger in the SC-TG experiment. The MS scans revealed that larger amounts of CO_2 were formed in the second reaction stage under TG conditions and metallic copper was identified in the reaction intermediate. Thus the higher concentration of gaseous reaction products under linear heating conditions resulted in the reaction:

$$Cu_2O + CO \rightarrow 2Cu + CO_2$$

This secondary reaction was avoided under sample controlled conditions where both the reaction temperatures and partial pressures of the gaseous reaction products were significantly lower.

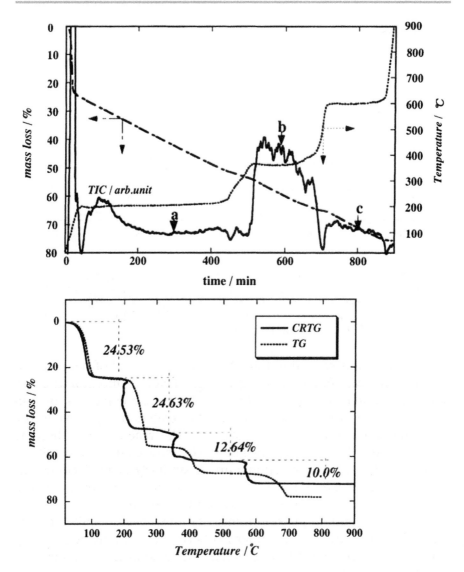

Figure 11.15 (a) Mass loss, temperature and total ion current (TIC) traces for simultaneous SC-TG/MS of calcium copper acetate hexahydrate at 0.06% min^{-1}; (b) comparison of mass loss traces for decomposition using a conventional TG of 3 °C min^{-1} with that using SC-TG at 0.06% min^{-1}.
Reprinted T. Arii and Y. Masuda, *Thermochim. Acta*, 1999, **342**, 139. Copyright (1999), with permission from Elsevier.[60]

11.4.7 Gas Concentration Programming

The gas concentration programming technique is illustrated by the reduction of copper(II) oxide by hydrogen to form metallic copper.

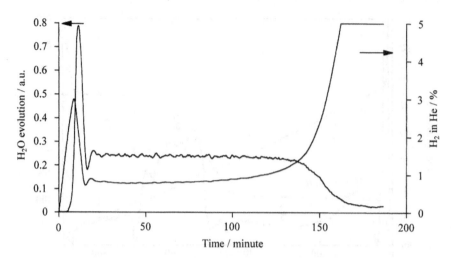

Figure 11.16 Constant rate reduction of 2.5 mg copper(ii) oxide at 225 °C min^{-1} with control on the hydrogen concentration. Reprinted from G. M. B. Parkes, P. A. Barnes and E. L. Charsley, *Thermochim. Acta*, 1998, **320**, 297. Copyright (1998), with permission from Elsevier.[61]

The reaction was carried out at 225 °C using a blend of hydrogen in helium and was followed using a hygrometer to monitor the formation of water during the reduction.[61] A typical curve is shown in Figure 11.16 and it can be seen that the reduction began when the hydrogen concentration was above 2%. The reaction then became very rapid and the hydrogen level had to be reduced to around 1.5% to maintain the selected reaction rate. A similarly shaped curve was observed in CR-EGA experiments in a hydrogen atmosphere and arises from the autocatalytic nature of the reduction.

Gas concentration programming has also been applied to the activation of a carbon char under isothermal conditions by blending air and nitrogen thus allowing considerable control of the development of porosity.[62]

11.4.8 Preparative Aspects

Many workers have realised the potential of SCTA (and CRTA in particular) as a preparative technique for finely divided and porous solids. The ability to control precisely the reaction rate offers a route to prepare homogenous materials with specified properties. This is illustrated by a CR-EGD study on the dehydration of gibbsite Al(OH)$_3$ to produce alumina.[63] When the dehydration was carried out under a residual water vapour pressure of 4 Pa, the maximum BET surface area of the product was only 40 $m^2 g^{-1}$, but at a residual pressure of 100 Pa, a tenfold increase in the surface area of the alumina was observed. The preparative aspects of SCTA for a range of materials including adsorbents, catalysts and ceramics have been discussed.[1,25]

SCTA also allows reaction intermediates to be prepared for study by X-ray diffraction and other analytical techniques, since it is easier to stop an experiment at a given extent of reaction than in linear heating experiments. This enables structural or other changes in a material to be more closely related to the extent of reaction. SCTA could also be used to determine the temperature profile required to obtain a particular product, which could then be approximated for use with a preparative scale system that lacks SCTA control.

11.5 Summary

The SCTA approach, where the heating rate is governed by the sample itself, offers a significant number of benefits over conventional thermal analysis. These include enhanced resolution, improved kinetic data and better preparative methods for finely divided and porous solids. The availability of SCTA software for many commercial instruments has now made the technique accessible to a large number of thermal analysis practitioners. However, the family of SCTA techniques should not be seen as replacements for linear heating methods but rather as valuable complementary techniques and their utilisation is to be encouraged.

Further Reading

A detailed coverage of the theory and application of SCTA to kinetics, ceramics, adsorbents and catalysts is given in the book *Sample Controlled Thermal Analysis* edited by O. T. Sørensen and J. Rouquerol.[1] Good overviews of SCTA are also given by Rouquerol[64] and by Reading[7] and its application to polymers is discussed by Sánchez-Jiménez *et al.*[21] A recent review of the application of CRTA to kinetics and to the synthesis of materials is provided by Pérez-Maqueda *et al.*[25] In addition to the cited references, the application notes available from the thermal analysis instrument manufacturers are a useful source of reference on the application of SCTA to different materials.

References

1. *Sample Controlled Thermal Analysis*, ed. O. T. Sørensen and J. Rouquerol, Kluwer Academic Publications, Dordrecht, 2003.
2. T. Lever, P. Haines, J. Rouquerol, E. L. Charsley, P. Van Eckeren and D. J. Burlett, *Pure Appl. Chem.*, 2014, **86**, 545.
3. J. Rouquerol, *Bull. Soc. Chim. Fr.*, 1964, 31.
4. L. Erdey, F. Paulik and J. Paulik, Patent, 1962 (registered) 1965 (published).
5. O. T. Sørensen, *J. Therm. Anal.*, 1978, **13**, 429.
6. E. L. Charsley, J. O. Hill, G. M. B. Parkes and J. J. Rooney, Poster Presented at 8th European Symposium on Thermal Analysis and Calorimetry, Barcelona, Spain, 2002.
7. M. Reading, in *Handbook of Thermal Analysis and Calorimetry*, ed. M. E. Brown, Elsevier, Amsterdam, 1998, ch. 8, vol. 1, pp. 423–443.
8. G. M. B. Parkes, P. A. Barnes and E. L. Charsley, *Anal. Chem.*, 1999, **71**, 2482.
9. G. Thevand, F. Rouquerol and J. Rouquerol, in *Thermal Analysis*, ed. B. Miller, John Wiley and Sons, New York, 1982, vol. 2, pp. 1524–1530.
10. O. T. Sørensen, *J. Therm. Anal.*, 1992, **38**, 213.
11. M. Reading, in *Thermal Analysis - Techniques and Applications*, eds. E. L. Charsley and S. B. Warrington, Royal Society of Chemistry, Cambridge, 1992, pp. 126–155.
12. P. S. Gill, S. R. Sauerbrunn and B. S. Crowe, *J. Therm. Anal.*, 1992, **38**, 255.
13. R. Riesen, *J. Therm. Anal. Calorim.*, 1998, **53**, 365.
14. E. Kapsch and G. Kaiser, *Netzsch Onset*, 2013, **11**, 9–11.
15. G. M. B. Parkes, P. A. Barnes, E. L. Charsley, M. Reading and I. Abrahams, *Thermochim. Acta*, 2000, **354**, 39.
16. E. A. Fesenko, P. A. Barnes, G. M. B. Parkes, E. A. Dawson and M. J. Tiernan, *Top. Catal.*, 2002, **19**, 283.
17. E. A. Dawson, G. M. B. Parkes, P. A. Barnes, M. J. Chinn, L. A. Pears and C. J. Hindmarsh, *Carbon*, 2002, **40**, 2897.
18. A. Ortega, *Int. J. Chem. Kinet.*, 2002, **34**, 223.
19. J. M. Criado, A. Ortega and F. Gotor, *Thermochim. Acta*, 1990, **157**, 171.
20. L. A. Pérez-Maqueda, A. Ortega and J. M. Criado, *Thermochim. Acta*, 1996, **277**, 165.
21. P. E. Sánchez-Jiménez, L. A. Pérez-Maqueda, A. Perejón and J. M. Criado, *Polym. Degrad. Stab.*, 2011, **96**, 974.

22. F. Rouquerol and J. Rouquerol, in *Thermal Analysis*, ed. H. G. Wiedemann, Birhhauser, Basel, 1972, vol. 1. pp. 373–377.
23. K. Nahdi, P. Llewellyn, F. Rouquerol, J. Rouquerol, N. K. Ariguib and M. T. Ayedi, *Thermochim. Acta*, 2002, **390**, 123.
24. J. M. Criado and L. A. Perez-Maqueda, in *Sample Controlled Thermal Analysis*, ed. O. T. Sørensen and J. Rouquerol, Kluwer Academic Publishers, Dordrecht, 2003, ch. 4.
25. L. A. Pérez-Maqueda, J. M. Criado, P. E. Sánchez-Jiménez and M. J. Diánez, *J. Therm. Anal. Calorim.*, 2015, **120**, 45.
26. F. Paulik, *Special Trends in Thermal Analysis*, John Wiley & Sons Ltd, Chichester, UK, 1995.
27. PerkinElmer, *Application Note 44-74045*, 2009.
28. T. Arii and N. Fujii, *J. Anal. Appl. Pyrolysis*, 1997, **39**, 129.
29. SETARAM Instrumentation, *Application Note AN097*.
30. E. L. Charsley, J. J. Rooney, H. A. White, B. Berger and T. T. Griffiths, in *Proceedings 31st North American Thermal Analysis Society Conference*, ed. M. Rich, NATAS, Albuquerque, New Mexico, 2003, p. 133.
31. M. J. Fernández-Berridi, N. González, A. Mugica and C. Bernicot, *Thermochim. Acta*, 2006, **444**, 65.
32. N. Gonzalez, A. Mugica and M. J. Fernandez-Berridi, *Polym. Degrad. Stab.*, 2006, **91**, 629.
33. C. Gamlin, M. G. Markovic, N. K. Dutta, N. R. Choudhury and J. G. Matisons, *J. Therm. Anal. Calorim.*, 2000, **59**, 319.
34. S. Renneckar, A. G. Zink-Sharp, T. C. Ward and W. G. Glasser, *J. Appl. Polym. Sci.*, 2004, **93**, 1484.
35. A. Zanier, *J. Therm. Anal. Calorim.*, 2001, **64**, 377.
36. M. A. Osman, M. Ploetze and U. W. Suter, *J. Mater. Chem.*, 2003, **13**, 2359.
37. J. I. Tobón, J. J. Payá, M. V. Borrachero and O. J. Restrepo, *Const. Build. Mater.*, 2012, **36**, 736.
38. T. Arii and A. Kishi, *J. Therm. Anal. Calorim.*, 2006, **83**, 253.
39. J. Rouquerol, *Thermochim. Acta*, 1997, **300**, 247.
40. V. Chevrot, P. L. Llewellyn, F. Rouquerol, J. Godlewski and J. Rouquerol, *Thermochim. Acta*, 2000, **360**, 77.
41. E. Badens, P. Llewellyn, J. M. Fulconis, C. Jourdan, S. Veesler, R. Boistelle and F. Rouquerol, *J. Solid State Chem.*, 1998, **139**, 37.
42. M. Reading and J. Rouquerol, *Thermochim. Acta*, 1985, **85**, 305.
43. J. Rouquerol, S. Bordère and F. Rouquerol, *Thermochim. Acta*, 1992, **203**, 193.
44. P. A. Barnes, G. M. B. Parkes, D. R. Brown and E. L. Charsley, *Thermochim. Acta*, 1995, **269–270**, 665.

45. O. T. Sørensen and J. M. Criado, in *Sample Controlled Thermal Analysis*, eds. O. T. Sorensen and J. Rouquerol, Kluwer Academic Publishers, Dordrecht, 2003, pp. 102–134.

46. O. T. Sørensen, *J. Therm. Anal. Calorim.*, 2003, **72**, 1093.

47. B. Paul, D. Jain, S. P. Chakraborty, I. G. Sharma, C. G. S. Pillai and A. K. Suri, *Thermochim. Acta*, 2011, **512**, 134.

48. A. Ghosh, S. Koley, A. K. Sahu, T. Kundu Roy, S. Ramanathan and G. P. Kothiyal, *J. Therm. Anal. Calorim.*, 2014, **115**, 1303.

49. H. Palmour and D. R. Johnson, in *Sintering & Related Phenomena*, ed. G. C. Kuczynski, N. A. Hooton and C. F. Gibbons, Gordon & Breach, New York, 1967, pp. 779-791.

50. SETARAM Instrumentation, *Application Note AN351*.

51. C. S. Smith, *Trans. AIME (Metals Division)*, 1940, **137**, 236.

52. A. Saccone, D. Macciò, J. A. J. Robinson, F. H. Hayes and R. Ferro, *J. Alloys Compd.*, 2001, **317–318**, 497.

53. F. Paulik, E. Bessenyey-Paulik and K. Walther-Paulik, *Thermochim. Acta*, 2003, **402**, 105.

54. F. Paulik, E. Bessenyey-Paulik and K. Walther-Paulik, *Thermochim. Acta*, 2005, **430**, 59–65.

55. E. L. Charsley, P. G. Laye, G. M. B. Parkes and J. J. Rooney, *J. Therm. Anal. Calorim.*, 2011, **105**, 699.

56. E. L. Charsley, C. Stewart, P. A. Barnes and G. M. B. Parkes, *J. Therm. Anal. Calorim.*, 2003, **72**, 1087.

57. J. Rouquerol, in *Thermal Analysis*, ed. R F. Schwenker and P. D. Garn, Academic Press, New York, 1969, vol. 1, pp. 281–288.

58. L. A. Pérez-Maqueda, J. M. Criado and F. J. Gotor, *Int. J. Chem. Kinet.*, 2002, **34**, 184.

59. N. Koga and S. Yamada, *Int. J. Chem. Kinet.*, 2005, **37**, 346.

60. T. Arii and Y. Masuda, *Thermochim. Acta*, 1999, **342**, 139.

61. G. M. B. Parkes, P. A. Barnes and E. L. Charsley, *Thermochim. Acta*, 1998, **320**, 297.

62. E. A. Dawson, G. M. B. Parkes, P. A. Barnes, M. J. Chinn and P. R. Norman, *Thermochim. Acta*, 1999, **335**, 141.

63. J. Rouquerol and M. Ganteaume, *J. Therm. Anal.*, 1977, **11**, 201.

64. J. Rouquerol, *J. Therm. Anal. Calorim.*, 2003, **72**, 1081.

Subject Index

Page numbers in *italics* indicate a figure; page numbers with T indicate a table.